Race and Membership in American History: The Eugenics Movement

Facing History and Ourselves National Foundation, Inc.
Brookline, Massachusetts

For permission to reproduce the following photographs, posters, and charts in this book, grateful acknowledgement is made to the following:

Cover: "Mixed Types of Uncivilized Peoples" from Truman State University. (Image #1028 from Cold Spring Harbor Eugenics Archive, *http://www.eugenics archive.org/eugenics/).* Fitter Family Contest winners, Kansas State Fair, from American Philosophical Society (image #94 at *http://www.amphilsoc.org/ library/guides/eugenics.htm*). Ellis Island image from the Library of Congress. Petrus Camper's illustration of "facial angles" from *The Works of the Late Professor Camper* by Thomas Cogan, M.D., London: Dilly, 1794.
Inside: p. 45: *The Works of the Late Professor Camper* by Thomas Cogan, M.D., London: Dilly, 1794. **51**: "Observations on the Size of the Brain in Various Races and Families of Man" by Samuel Morton. *Proceedings of the Academy of Natural Sciences*, vol. 4, 1849. **74**: The American Philosophical Society. **77**: *Heredity in Relation to Eugenics*, Charles Davenport. New York: Henry Holt &Co., 1911. **99**: Special Collections and Preservation Division, Chicago Public Library. **116**: The Missouri Historical Society. **119**: *The Daughters of Edward Darley Boit, 1882*; John Singer Sargent, American (1856-1925). Oil on canvas; 87 3/8 x 87 5/8 in. (221.9 x 222.6 cm.). Gift of Mary Louisa Boit, Julia Overing Boit, Jane Hubbard Boit, and Florence D. Boit in memory of their father, Edward Darley Boit, 19.124. Museum of Fine Arts, Boston. **121**: *America and Lewis Hine, Photographs 1904-1940,* 1977. Aperture, Inc. **143**: The American Philosophical Society. **144**: Reprinted with permission of the American Philosophical Society. (AES Scrapbook #44.) **145**: The Ohio Historical Society. **157, 158**: *Memoirs of the National Academy of Sciences*, vol. XV, "Psychological Examining in the United States Army," Washington: GPO, 1921. **160**: Brigham, Carl. *A Study of American Intelligence.* Copyright © 1923 by Princeton University Press. Reprinted by permission of Princeton University Press. **168**: American School Board Journal, 1922. **197**: Truman State University. **212**: The Library of Congress. **223**: The American Philosophical Society. **245**: Cold Spring Harbor (image #672). **247**: *Racial Hygiene: Medicine Under the Nazis* by Robert Proctor. Cambridge, Ma.: Harvard Universtiy Press, 1988. **260**: Deutsches Hygiene-Museum. **264**: *Der Sturmer*, April 15, 1943. **278**: *Volk und Rasse*, 1933.

Acknowledgement of the many documents and other quotations included in this book may be found at the end of each reading. Every effort has been made to trace and acknowledge owners of copyrighted materials, but in some cases that has proved impossible. Facing History and Ourselves would be pleased to add, correct, or revise such citations in future printings.

Facing History and Ourselves, 16 Hurd Road, Brookline, MA 02445
(617) 232-1595 *www.facinghistory.org*

Printed in the United States of America ISBN 0-9615841-9-K

IN DEDICATION

Race and Membership in American History: The Eugenics Movement is dedicated to Seth A. Klarman, chair of the Facing History and Ourselves Board of Directors and president of The Baupost Group, L.L.C., with the utmost gratitude for his leadership, wisdom, and kindness.

As Seth's partners at Baupost, we considered many ways in which we could honor Seth and express our thanks to him for building such a successful business. We could think of no better way than highlighting the role he has played in leading Facing History and Ourselves. This critical organization has been so important to Seth's personal exploration, both intellectual and practical, of social responsibility. Seth, in turn, has devoted his boundless energy and curiosity, good counsel and generosity to helping make Facing History a success. We chose to dedicate this book to Seth because it exemplifies his deep interest in history, his commitment to intellectual rigor, and his strong conviction about the importance of Facing History's mission—and not least, because he has read and reviewed every word in draft form.

Seth and his wife Beth, who chairs the New England Regional Advisory Board, are vital members of the Facing History family. They have dedicated their time and philanthropy to the organization because they know that Facing History addresses some of the most difficult issues facing our society—the hatreds that exist in our world and how we confront them. Together, their dedication has made a difference to the work of Facing History.

We thank Seth for his hard work on our behalf, the high standard he sets, and his vision that has helped build successful organizations. His ability to combine the best in philanthropy and business with a commitment to family sets a model and helps make the world a better place.

With sincere thanks and true friendship,

Thomas W. Blumenthal
Paul C. Gannon
Scott A. Nathan
Samuel Plimpton

CONTENTS

FOREWORD

Race and Membership in American History: The Eugenics Movement has been a long time in coming. After years of revising drafts, piloting lessons in classrooms, and infusing ideas into institutes, we now have a special book that I believe was worth the wait.

The book asks us to rethink what we know about our own past. While barely remembered today, the eugenics movement represents a moral fault line in our history. It was a movement that defined differences in terms of racially superior and inferior human traits. Because these ideas were promoted in the name of science and education, they had a dramatic impact on public policies and the lives of ordinary people at the time and, in turn, created legacies that are still with us today. The eugenics movement is not a historical footnote. It is a fundamental chapter in our history that ought to be examined in our classrooms.

When I first began research on eugenics in 1993, I sensed there was an important story to be told. That story became a manuscript that I completed in 1997. It is now a book that has been rewritten, edited, and commented upon by many people. Often overlooked are those who volunteered their time to comb archives and correspond with scholars. I will not forget the research efforts of Debbie Karpel, Lisa Rivo, Charlie Putnam, Kirsten Greenidge, and Lisa Middents. I thank them very much.

I would like to acknowledge the scholars who have taught me so much about the forgotten history of eugenics and who have also taken the time to review drafts of this book. First is Paul Lombardo who has so kindly sent us his pioneering articles on the history of sterilization and antimiscegenation laws. He has advised us when called upon and recommended our program to educators around the country. I very much respect his intellect and his keen ethical sense of justice in the work he does.

My appreciation is extended to Steven Selden who first alerted me to the role eugenics has played in curriculum and school organization. His book, *Inheriting Shame: The Story of Eugenics and Racism in America*, is a must-read for anyone who wants to see how eugenic ideas became part of school curricula. I also want to thank Nicole Hahn Rafter for her special contributions to understanding eugenic conceptions of the criminal. Her books, *White Trash: The Eugenic Family Studies* and *Creating Born Criminals* were extremely valuable in helping us to think about how notions of the "criminal other" have become institutionalized over time. Finally, I want to thank Facing History and Ourselves for its support of the book and its mission to provide education for democracy.

Alan Stoskopf
Associate Program Director for Staff Development

PREFACE

Since its inception in 1976, Facing History and Ourselves has been exploring the roots of racism and antisemitism. Eugenics—an early 20th century movement to rid society of "inferior racial traits"—is part of that history. I learned about the movement from Stephan Chorover, the father of one of my first Facing History students. In *From Genesis to Genocide,* he described the connections between the history of "race science" and eugenics in the United States and Nazi programs that aimed at "purifying the race" by murdering millions of children, women, and men.

As Facing History's interest in the eugenics movement deepened, we featured it at our 1992 conference, "Examining Historical Roots to Racism and Antisemitism: A Profile of Facing History's Research" and invited Stephen Jay Gould, who has written about the movement, to be the featured speaker. As we prepared for the event, we studied Gould's work, particularly his now classic *Mismeasure of Man.* In the introduction, he describes a conversation that appears in Plato's *Republic*. Socrates explains to Glaucon that in his ideal society citizens will be assigned to one of three classes. Uncertain as to how he will persuade people to accept such a division, Socrates decides to create a myth.

> Citizens, we shall say to them in our tale, you are brothers, yet God has framed you differently. Some of you have the power of command, and in the composition of these he has mingled gold, wherefore also they have the greatest honor; others he has made of silver, to be auxiliaries; others again who are to be husbandmen and craftsmen he has composed of brass and iron; and the species will generally be preserved in the children.[1]

When asked if citizens will believe the myth, Glaucon replies, "Not in the present generation . . . but their sons may be made to believe in the tale, and their son's sons, and posterity after them." That myth is still being taught, studied, and believed. Gould's book explores the scientific version of the myth. In *Killers of the Dream,* Lillian Smith, a white southerner, tells of how that same myth shaped her childhood in the early 1900s:

> The mother who taught me what I know of tenderness and love and compassion taught me also the bleak rituals of keeping Negroes in their "place." The father who rebuked me for an air of superiority toward schoolmates from the mill and rounded out his rebuke by gravely reminding me that "all men are brothers," trained me in the steel-rigid decorums I must demand of every colored male. . . .
>
> From the day I was born, I began to learn my lessons. . . . I learned it is possible. . . to pray at night and ride a Jim Crow car the next morning and to feel comfortable doing both. I learned to believe in freedom, to glow when the word democracy was used, and to practice slavery from morning to night. I learned it the way all of my southern people learn it: by closing door after

door until one's mind and heart and conscience are blocked off from each other and from reality.[2]

Eventually Smith's struggle with her conscience led her to publicly challenge the myth. *Race and Membership in American History: The Eugenics Movement* also confronts that myth by describing the support it received from a twisted science that betrayed generations of children and turned neighbor against neighbor.

This new resource book reflects a decade of research and development inspired in part by the teaching of K. Anthony Appiah and funded through the Harvard/Facing History and Ourselves Project. Alan Stoskopf wrote the original manuscript, introduced our staff to leading scholars in the field, shared findings at board meetings, institutes, and workshops in the United States and Europe, and built a team of colleagues who reviewed and piloted lessons in dozens of Facing History classrooms. It is a contribution we value greatly.

Under the leadership of Marc Skvirsky, the staff has incorporated our research into Facing History programs and materials and made it a part of our professional development. Among those who assisted in these efforts are Jennifer Clark, Kevin Feinberg, Karen Murphy, and Adam Strom. Phyllis Goldstein integrated the outcome of our mutual efforts into a resource book with the assistance of Tracy O'Brien and Karen Lempert, who prepared bibliographies and secured permissions, and Jenifer Snow, the book's designer. John Englander has relied on that research in creating an instructional module on the eugenics movement for our website at *www.facinghistory.org.*

Both the resource book and the module reveal the power of ideas to shape reality and the importance of education in preserving and protecting democracy. Scientist Jacob Bronowski, who lost members of his own family in the Holocaust, believed that societies are held together by mutual respect. He has written that a society fails, "in fact, it falls apart into groups of fear and power, when the concept of man is false." In his view, the never-ending search for what makes us human helps keep democracy alive. To turn scientific inquiry into dogma is in Bronowski's words to "freeze the concept of man into a caricature beyond correction, as the society of caste and master race have done." Exploring our common humanity by confronting the myth of superior "races," classes, and individuals is essential to education for democracy in the 21st century. History suggests we have a lot to learn.

Margot Stern Strom
Executive Director

---

1 Quoted in *Mismeasure of Man* by Stephen Jay Gould. W.W. Norton & Company, 1996, 1981, pp. 51-52.

2 *Killers of the Dream* by Lillian Smith. W.W. Norton & Company, 1949. Doubleday Anchor edition, 1963, pp. 17, 18-19.

OVERVIEW

R*ace and Membership in American History: The Eugenics Movement* focuses on a time in the early 1900s when many people believed that some "races," classes, and individuals were superior to others. They used a new branch of scientific inquiry known as eugenics to justify their prejudices and advocate programs and policies aimed at solving the nation's problems by ridding society of "inferior racial traits."

Rationale

Eugenics was an international movement that attracted heads of state, teachers, philanthropists, journalists, and ordinary citizens who advocated laws and policies that would shape the most basic decisions that individuals and societies make: Who may marry? Who may have children? Who will be educated? Who may live among us?

Race and Membership in American History considers how people responded to those questions at various times in history. The book takes on special importance at a time when scientists have just completed the first survey of the human genome—a scientific milestone that promises to enhance our understanding of the ways inherited traits influence who we are and what we may become. There is much we as citizens can learn about the relationship between science and society from the history of the eugenics movement.

In Nazi Germany, eugenics was used to shape and ultimately justify policies of mass murder. In the United States, the consequences were less extreme. Nonetheless, eugenics had a profound effect on almost every aspect of everyday life. Long after other scientists had shown that the laws of heredity are more complicated than "breeding the best with the best," eugenicists were still trying to segregate and sterilize the mentally and physically disabled. Long after anthropologists had shown that intelligence and other human traits are shaped at least in part by culture and environment, eugenicists were still seeking ways to "protect" the purity of the "white race."

Links to the Curriculum

Race and Membership can be used as a companion to Facing History's primary resource book, *Holocaust and Human Behavior.* It may also be used to deepen a study of American or world history. The book's vast array of primary sources can also provide a rich historical context for courses in literature, civics, education, psychology, sociology, and government. In addition, the book can be used to enhance a study of the history of science.

Organization

Like other Facing History publications, *Race and Membership in American History* is a resource book that provides a flexible structure for examining complex events and ideas. Teachers are encouraged to select the readings that are most appropriate for their students and which best match the objectives of their curriculum. They are also encouraged to choose the questions and activities that further those objectives from "Connections," a section at the end of every reading with suggestions for discussion, writing assignments, and research projects.

Race and Membership in American History is a departure from other Facing History resource books in that, for the first time, a number of related readings appear on the web at *www.facinghistory.org* as part of an instructional module that traces the connections between the American eugenics movement and its counterpart in Nazi Germany.

Scope and Sequence

Like all Facing History publications, this new resource book begins with questions of identity and membership. It then moves to a study of a particular history—one that fosters an understanding of the role of citizen, the fragility of democracy, the ways prejudices and other preconceived notions can distort scientific inquiry, and the dangers of resolving complex problems by dividing the world into *us* and *them* and then blaming "them" for all of the ills of society. It is a history that also raises profound questions of right and wrong, of guilt and responsibility. As in other Facing History publications, a variety of questions and activities link that history to our own lives and experiences. The book ends with a thoughtful look not only at the legacies of this history but also of concerns related to prevention.

The book is divided into nine chapters. Chapter 1 introduces students to key concepts and themes by examining the idea of difference through various lenses. Chapter 2 places those ideas in historical perspective by examining how Europeans and Americans regarded differences in the 1700s and early 1800s. Many of the beliefs we hold today about race, citizenship, and democracy developed during those years. The chapter focuses on the tension between two contradictory notions about human worth—racism and equality—the consequences of that tension, and its effects on the lives of real people in the past and today.

Chapter 3 brings notions about human worth into the 20th century by describing the origins of the eugenics movement. Chapter 4 places that movement in a historical context by examining its links to the progressive movement. Chapters 5, 6, and 7 consider how eugenicists used science to justify social inequalities, deny opportunities, and legitimize violence. The chapters also offer insights into the fragility of democracy by exploring why the movement was attractive to many Americans in the early 1900s.

Chapter 8 outlines the connections between the American and German eugenics movements. Those links reveal fault lines in the relationship between science and society. In Chapter 9, students move from a study of the past to questions of judgment and participation. The first two readings in this chapter return to the questions of Chapter 1. The readings that follow apply those questions to current discussions of the relationship between science and society. Each of these readings is followed by suggestions for independent research or group projects. The book ends with "For Further Investigation," a list of websites, books, videos, and other resources.

1. Science Fictions and Social Realities

To be a difference, a difference has to make a difference.

Gertrude Stein

Who are you? It is a question that we have all been asked. In answering, we define ourselves by placing greater emphasis on some characteristics than on others. Most of us view our identity as a combination of many factors, including physical characteristics and social ties—connections to a family, an ethnic group, a community, a religion, or a nation. Although this way of defining a person seems ordinary, it has consequences. "When we identify one thing as unlike the others," writes Martha Minow, a law professor, "we are dividing the world; we use our language to exclude, to distinguish, to discriminate." She continues:

> Of course, there are "real differences" in the world; each person differs in countless ways from each other person. But when we simplify and sort, we focus on some traits rather than others, and we assign consequences to the presence and absence of the traits we make significant. We ask, 'What's the new baby?'—and we expect as an answer, boy or girl. That answer, for most of history, has spelled consequences for the roles and opportunities available to that individual.[1]

At what point do physical differences become powerful social divisions that affect what we believe is possible for others and ourselves? How are such differences used to justify social inequalities? What role do scientists, educators, religious leaders, and the media play in the process? How does history shape the value we place on *us* and *them*? This book explores how such questions were answered at specific times in American history. Chapter 1 introduces these questions by examining the idea of difference through various lenses.

In every society some differences matter more than they do in others. The way a society responds to differences affects the way individuals see themselves and others. Those responses are especially important at a time when scientists are completing "the first survey of the entire human genome"—a scientific milestone that promises to enhance our understanding of the ways inherited traits influence who we are and what we become. The readings in this chapter raise important questions about the relationship between our genetic inheritance and our identity. In doing so, they increase our awareness of the factors that shape not only how we see ourselves and others but also the value we place on our observations.

---

1. *Making All the Difference: Inclusion, Exclusion and American Law* by Martha Minow. Cornell University Press, 1990, p. 3.

What Do We Do with a Variation?

Reading 1

James Berry raises important questions about the ways we respond to differences in a poem entitled "What Do We Do with a Variation?"

> What do we do with a difference?
> Do we stand and discuss its oddity
> or do we ignore it?
>
> Do we shut our eyes to it
> or poke it with a stick?
> Do we clobber it to death?
>
> Do we move around it in rage
> and enlist the rage of others?
> Do we will it to go away?
>
> Do we look at it in awe
> or purely in wonderment?
> Do we work for it to disappear?
>
> Do we pass it stealthily
> Or change route away from it?
> Do we will it to become like ourselves?
>
> What do we do with a difference?
> Do we communicate to it,
> let application acknowledge it
> for barriers to fall down?[1]

CONNECTIONS

What is the message of Berry's poem? Imagine a poem that described how we are alike. What might the message of such a poem be?

Noy Chou, a high school student who was born in Cambodia and reared in the United States wrote the following stanza as part of a poem entitled "You Have to Live in Somebody Else's Country to Understand." It explains how she feels about being perceived as different.

> What is it like to be an outsider?
> What is it like to sit in the class where everyone has blond hair
> and you have black hair?
> What is it like when the teacher says, "Whoever wasn't born
> here raise your hand."
> And you are the only one.
> Then, when you raise your hand, everybody looks at you and
> makes fun of you.
> You have to live in somebody else's country to understand.[2]

What does the student add to your understanding of James Berry's question? How would you answer his question?

How did you learn which differences matter and which do not? Record your response in a journal. A journal can be a way of documenting the process of thinking. For author Joan Didion and others, it is also a way of examining ideas. She explains, "I write entirely to find out what I'm thinking, what I'm looking at, what I see and what it means." You may find it helpful to use a journal to explore not only the ideas raised in this chapter but also those in the chapters that follow.

In the *Guide to Choosing to Participate*, available from Facing History and Ourselves, Jesus Colón describes how real and perceived differences shaped a decision he made on a late-night subway ride in New York City in the 1950s. A young white woman on the train was struggling with two small children, a baby, and a suitcase. Colón wanted to help her but feared her response. Would she accept his help? Or would she see him as a threat because he was black and Puerto Rican? Colón called his essay "Little Things Are Big." What do you think the title of Colón's essay means? How does the title relate to the question Barry asks in his poem?

---

1. "What Do We Do with a Variation?" from *When I Dance* by James Berry, Harcourt Brace. Reprinted by permission of Harcourt, Inc.
2. "You Have to Live in Somebody Else's Country to Understand" by Noy Chou. In *A World of Difference Teacher/Student Guide,* Anti-Defamation League of B'nai B'rith and Facing History and Ourselves, 1986, Group and Individual Identity, pp. 8-9.

Eye of the Beholder

Reading 2

The Twilight Zone, a popular TV show from 1959 through 1965, blended science fiction with fantasy and horror. The action often took place in familiar settings and featured characters that seemed quite ordinary. Their stories, however, were far from ordinary because they lived in an imaginary world just beyond our own—"the twilight zone." In creating the series, producer and writer Rod Serling hoped it would prompt thoughtful discussions of social issues. "Eye of the Beholder," one of Serling's most provocative episodes, offers an answer to James Berry's question: What do you do with a difference? (Page 2).

A video of the episode (22 minutes) is available at video stores or may be borrowed from the Facing History Resource Center. If possible, watch the episode as a class. If you are unable to see it, the following paragraphs provide a synopsis of the story.

> Meet the patient in room 307, Janet Tyler. A rigid mask of gauze bandages covers her face. Only her voice and her hands seem alive as she pleads with a nurse to describe the weather, the sky, the daylight, clouds—none of which she can see. The nurse, visible only by her hands, answers kindly but briefly.
>
> "When will they take the bandages off?" Janet asks urgently. "How much longer?"
>
> "When they decide they can fix your face," the nurse replies.
>
> "It's pretty bad, isn't it? Ever since I was little, people have turned away when they looked at me. . . . The very first thing I can remember is another little child screaming when she saw me. I never wanted to be beautiful, to look like a painting. I just wanted people not to turn away."
>
> With a consoling pat, the nurse moves away.
>
> A doctor enters Janet Tyler's room. We see only his hands, his shadow, his back as he looks out a window. Janet questions him with a mixture of fear and hope. When will he remove the bandages? Will her face be normal?
>
> The doctor tries to comfort her. His voice is gentle. Perhaps this time the treatment will be successful. But he also issues a warning. He reminds her that she has had treatment after treatment—eleven in all. That is the limit. If this effort fails, she can have no more.
>
> "Each of us is afforded as much opportunity as possible to fit in with society," he says. "In your case, think of the time and effort the

state has expended, to make you look—"

"To look like what, doctor?"

"Well, to look normal, the way you'd like to look. . . . You know, there are many others who share your misfortune, who look much as you do. One of the alternatives, just in case the treatment is not successful, is to allow you to move into a special area in which people of your kind have congregated."

Janet twists away from the doctor. "People of my kind? Congregated? You mean segregated! You mean imprisoned! You are talking about a ghetto—a ghetto for freaks!" Her voice rises in a crescendo of anger.

"Miss Tyler!" the doctor remonstrates sharply. "You're not being rational. You know you couldn't live any kind of life among normal people." His words are harsh, but his voice is sad and patient.

Janet refuses to be mollified. "Who are these normal people?" she asks accusingly. "Who decides what is normal? Who is this state that makes these rules? The state is not God! The state does not have the right to make ugliness a crime. . . . Please," she begs. "Please take off the bandages. Please take them off! Please help me."

Reluctantly the doctor agrees, and the staff prepares for the removal. Bit by bit, he peels the gauze away. She sees at first only the light, then the shadowy forms of the doctor and nurses. As the last strip of gauze comes off, the doctor and nurses draw back in dismay. "No change!" the doctor exclaims. "No change at all!"

Janet Tyler gasps and raises her face. She has wide-set eyes, delicate brows, fine skin, and regular features, framed by wavy blonde hair. She begins to sob and struggle away from the nurses.

"Turn on the lights," the doctor orders. "Needle, please!"

As the lights come on, the doctor and nurses are clearly visible for the first time. Piglike snouts dominate their lopsided, misshapen features. Their mouths are twisted, their jowls sag.

Janet runs through the hospital in a panic, pursued by nurses and orderlies. She passes other staff and patients. Each face is a little different but all share the same basic pattern—snouts, jowls, and all. She flings open a door and freezes in sudden shock. The doctor and another man are in this room. She sinks down by a chair and hides her face in fear.

"Miss Tyler, Miss Tyler, don't be afraid," the doctor urges. "He's only a representative of the group you are going to live with. He won't hurt you. . . . Miss Tyler, this is Walter Smith."

Walter Smith steps forward, and Janet Tyler cringes away. He too has regular features, lit by a friendly smile. A stray lock of dark hair curls over his forehead. "We have a lovely village and wonderful people," he tells Janet. "In a little while, a very little while, you'll feel a sense of great belonging."

She looks at his face. "Why do we have to look like this?" she murmurs.

"I don't know, I really don't," he replies with sadness. "But there is a very old saying—beauty is in the eye of the beholder. Try to think of that, Miss Tyler. Say it over and over to yourself. Beauty is in the eye of the beholder."

He holds out his hand to her. Slowly, hesitantly, she takes it, and they walk away together, through a corridor lined with pig-faced spectators.[1]

CONNECTIONS

List the words and phrases in the episode that you found significant. Be sure to identify the person who utters those words. (For example: Tyler: "Who decides what is normal?") What does your list suggest about the way difference is understood in this society?

Who in Janet Tyler's society determines what is "normal"? Who is "beautiful"? What is "rational"? What is the source of that power? Why is "ugliness" a crime? While this show is fiction, it raises important questions about images in our own society. Where do we get our ideas about beauty? How do we learn what is "normal"? What part does our family play? Our peers? What is the role of the media? To what extent do media images shape our standards of beauty? To what extent do those images reflect the views of society as a whole?

Our standards of beauty, ideas about difference, even notions of what is normal are shaped to a large extent by culture—the attitudes, values, and beliefs of a society. To find out how standards of beauty have changed over the years, you may wish to check movies made in the 20th century. Works of art can also offer clues to standards. So can toys—particularly dolls. For example, what do "Barbie" dolls suggest about our standards of beauty? What does your research suggest about the idea that "beauty is in the eye of the beholder"? About the way standards change? What events or ideas may have prompted those changes?

Describe the relationship between Janet Tyler and the doctor. Why does the doctor seem to have so much power and Tyler so little? Who do you think has power over the doctor?

How do the people in Tyler's society answer the question: What do you do with a difference? Where do you think people in that society got their ideas about difference? How do they learn which differences matter and which do not? Where do you get your ideas about difference? How do you learn which differences matter and which do not?

What part do the labels people place on differences—"disabled," "dysfunctional," "abnormal"—play in the way we view ourselves and others? According to an old saying, "Sticks and stones can break my bones but names can never hurt me." Is it true? Are labels harmless? Do words hurt?

Medicine is generally viewed as a healing profession and science as a body of knowledge that advances society. What is being "healed" in this society? How is society being "advanced"? What does the episode suggest about the relationship between physicians and other scientists and the society in which they live? For example, what does the episode suggest about the way physicians and scientists promote the values of their society? What does it suggest about the way the values of the larger society influence their work?

In the late 1950s and early 1960s, the television networks tried to avoid controversial issues. In a 1959 interview, Serling stated, "I think it's criminal that we are not permitted to make dramatic note of social evils that exist, of controversial themes, as they are inherent in our society. I think it's ridiculous that drama, which by its very nature should make a comment on those things which affect our daily lives, is in a position, at least in terms of television, of not being able to take a stand."[2] To what extent does Serling's criticism of television in the 1950s and 1960s hold true today? What "social evils" do TV dramas confront today? What "evils" are seldom discussed? For more information about the producer and the series that he created, see *Serling: The Rise and Twilight of Television's Last Angry Man* by Gordon Sander.

Because Serling's programs were science fiction, he had more freedom to deal with the issues of social injustice. To what social inequalities might this episode refer? How would you adapt the "Eye of the Beholder" for today's world? What changes would you make in the story?

---

1. Adapted from "Eye of the Beholder," (Nov. 11, 1960) *The Twilight Zone* and *The Twilight Zone Companion* by Marc Scott Zicree, 2nd Edition. Bantam Books, 1982, 1989, pp.144–145.
2. *The Mike Wallace Show*, Rod Serling, October 1, 1959, CBS.

Beyond Classification

Reading 3

Rod Serling used fiction to explore the negative consequences of the labels we attach to ourselves and others. For many Americans, that kind of discrimination is a part of their daily life. Stereotyping obscures the reality of who they are and what they may become.

According to many psychologists, although it is natural to view others as representatives of groups, stereotypes are offensive. They are more than a label or judgment about an individual based on the characteristics of a group. Stereotyping reduces individuals to categories. Therefore stereotyping can lead to prejudice and discrimination. The word *prejudice* means pre-judge. We prejudge when we have an opinion about a person based on his or her membership in a particular group. A prejudice attaches value to differences to the benefit of one's own group and at the expense of other groups. Discrimination occurs when prejudices are translated into actions. Not every stereotype results in discrimination. But all stereotypes tend to divide a society into *us* and *them.*

Dalton Conley understands the power of stereotypes. He writes:

> I am not your typical middle-class white male. I am middle class, despite the fact that my parents had no money; I am white, but I grew up in an inner-city housing project where most everyone was black or Hispanic. I enjoyed a range of privileges that were denied my neighbors but that most Americans take for granted. In fact, my childhood was like a social science experiment: Find out what being middle class really means by raising a kid from a so-called good family in a so-called bad neighborhood. Define whiteness by putting a light-skinned kid in the midst of a community of color. If the exception proves the rule, I'm that exception.
>
> Ask any African American to list the adjectives that describe them and they will likely put black or African American at the top of the list. Ask someone of European descent the same question and white will be far down the list, if it's there at all. Not so for me. I've studied whiteness the way I would a foreign language. I know its grammar, its parts of speech; I know the subtleties of its idioms, its vernacular words and phrases to which the native speaker has never given a second thought. There's an old saying that you never really know your own language until you study another. It's the same with race and class.

> In fact, race and class are nothing more than a set of stories we tell ourselves to get through the world, to organize our reality. . . . One of [my mother's favorite stories] was how I had wanted a baby sister so badly that I kidnapped a black child in the playground of the housing complex. She told this story each time my real sister, Alexandra, and I were standing, arms crossed, facing away from each other after some squabble or fistfight. The moral of the story for my mother was that I should love my sister, since I had wanted to have her so desperately. The message I took away, however, was one of race. I was fascinated that I could have been oblivious to something that years later feels so natural, so innate as race does.[1]

Diana Chang was born in New York City and reared in China. After returning to the United States, she wrote a poem called "Saying Yes."

> "Are you Chinese?"
> "Yes."
>
> "American?"
> "Yes."
>
> "Really Chinese?"
> "No . . . not quite."
>
> "Really American?"
> "Well, actually, you see . . ."
>
> But I would rather say
> yes
>
> Not neither-nor,
> not maybe,
> but both, and not only
>
> The homes I've had,
> the ways I am
>
> I'd rather say
> twice,
> yes.[2]

Ifemoa J. Nwokoye has lived in the United States and Nigeria. Her mother is a white American and her father a Nigerian. In both nations, people regard her as "different." She writes:

In our society, being both black and white is a difficult thing to deal with; you learn from the beginning that you are supposed to be a member of some specific group and so will never be accepted for who you really are. You are born into a complex world that aims to simplify things by making divisions between races. In America, people are often unwilling to accept the idea of a biracial person. In our everyday lives we are constantly confronted with situations in which we must define who we are. We check the boxes marked "white," "black," on our college forms, but there is no space marked "multiracial" yet. There is no place for me.

It is also twice as hard coming from two very distinct cultures—Nigerian and American. In each society I am treated in extremely different ways; yet, in both, I am identified by color. In America, I'm seen as black. I remember the time a schoolmate asked a friend of mine why she was sharing her snack with a black girl. I recall the icy stares of the ladies behind the perfume and make-up counters of every department store, their plastic smiles melting to frowns as they watched my every move. Most vividly, however, I remember how my math teacher would repeatedly confuse me with the only other black girl in the class, even until the end of the year—his belief apparently being that all black people look alike. Through all my experiences living in this culture, it has been a struggle to maintain my self worth.

Ironically, in Nigeria the situation is absolutely reversed. Because I am so much lighter than most people there, I am given a higher status and considered a model for others. I am treated with the utmost respect and admiration because in their eyes, I resemble a white person. What does remain consistent in both cultures is that I am not considered a biracial person; I'm still being labeled as one or the other.

I lived in Nigeria for the first seven years of my life and have visited on and off since my parents' divorce. As a child in Nigeria, I wasn't fully aware of people's perceptions of me, but I had a sense that I was somehow "better" than most of the children I knew, and that I had something special that they lacked. I remember being the teacher's favorite; the other students would get beaten, while I never experienced a lash of my teacher's cane. And I recall sitting in the front seat of my dad's car during a traffic jam. The little hands and noses of the village children would press hard against the window of the car, as if to penetrate the barrier of glass to steal a precious part of me. The society conditioned me to view myself as superior.

Drawing on my experience in America and in Nigeria, I have

reluctantly come to the conclusion that there is no place in either of my cultures where I can be accepted for who I am. I think of the irony in both experiences, and I don't know whether to laugh or cry.

I know that I must ignore the limitations and labels society places on me, and instead, realize that I am an individual with unique insight, able to encompass the best of both worlds. I refuse to see my biracial identity as confining, and I am determined not to be defeated by other people's narrow vision. Increasingly I am able to get strength from my inner voice and accept my own perspective on who I am. I now take pride in my two cultures.[3]

CONNECTIONS

Create an identity chart for Dalton Conley. The diagram below is an example of such a chart. It contains the words or phrases people attach to themselves as well as the ones that society gives them. Begin with the words or phrases that he uses to describe himself. Then add the labels others might attach to him. Create a similar chart for Ifemoa J. Nwokoye. What words or phrases does she use to describe herself? What words or phrases might others use to describe her? How are the two charts alike? What differences seem most striking? What part have labels played in shaping each identity?

Construct an identity chart for yourself, much like the ones you made for Conley and Nwokoye. After you have completed your chart, compare it to those

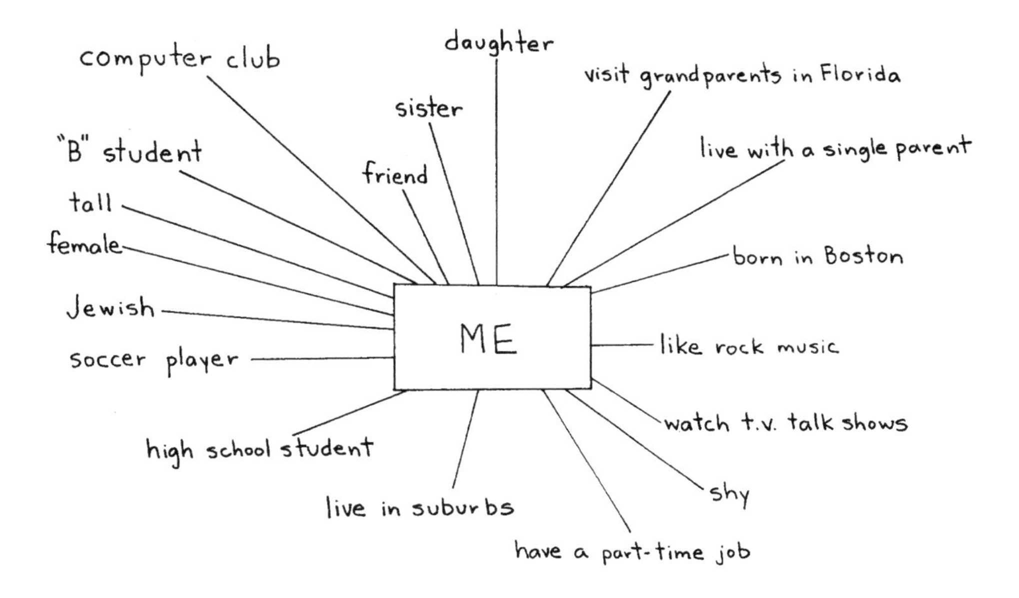

of your classmates. What do you have in common? To what extent is each chart unique? What part have labels played in shaping your identity? What other factors have affected who you are and what you may become?

Our identity—our sense of who we are and what we might become—is more than a set of labels. It is also shaped by our philosophy—our ideas, values, and beliefs about the world and our place in it. What ideas about self and others have shaped the philosophy of each of the individuals quoted in this reading?

How does society shape our identity? To what extent does your answer explain why Ifemoa Nwokoye believes that the identity crisis everyone faces is "doubly hard" for biracial children? Who else in American society may confront similar challenges in forging an identity?

Compare and contrast Diana Chang's experiences with those of Ifemoa Nwokoye. How do their histories shape their attitudes and beliefs?

A Facing History and Ourselves student who dislikes being labeled was surprised to discover that her classmates had similar feelings. She writes:

> I had always known that I didn't fit into boxes and labels neatly, but it was not until all of us in class looked carefully at our identities that I realized that there were times when we all couldn't fit into a box: racially, economically, religiously, or politically. That day we put away facades, superficial stereotypes, and imposed labels and came to the understanding that we are all crossbreeds in some way. . . . Once we were able to understand our own identities, we were better able to understand those of others.

How do the student's comments explain why Dalton Conley believes that "race and class are nothing more than a set of stories we tell ourselves to get through the world, to organize our reality"? How do the comments explain why Ifemoa Nwokoye has come to believe that she must "ignore the limitations and labels society places on me, and instead, realize that I am an individual with unique insight, able to encompass the best of both worlds"? How do the student's comments explain the way Nwokoye views her heritage?

---

1. *Honky* by Dalton Conley. University of California Press, 2000, pp. xi–xii.
2. Copyright © 1985 by Diana Chang.
3. *Of Many Colors: Portraits of Multiracial Families* was copyrighted ©1997 by Peggy Gillespie and was published in 1998 by the University of Massachusetts Press. pp. 137-138.

Reading 4

The difficulties many Americans have with labels are reflected in their responses to the nation's census. Every ten years, the United States government conducts a count of people living in the nation. A census is more than a count, however. It is a statistical portrait of the nation that provides detailed information about who we are and how we live. Every census has asked about race and every census has defined race differently.

As the 2000 census approached, many Americans urged the government to abandon questions dealing with race. Others favored the idea of adding a new box to the census form labeled "Multiracial." The government responded to the criticism by allowing individuals to check more than one "racial" box. After receiving his census questionnaire, journalist Courtland Milloy of the *Washington Post* wrote:

> A question on my U.S. Census survey asked: What is your race?
>
> The possible answers have been expanded this year to 17 and include space to write in "some other race," such as "cablinasian," as golfer Tiger Woods likes to call himself.
>
> A *Post* colleague, who is white, said he was going to check the black box—just for the hell of it, I suppose.
>
> "What are they going to do, put me in jail?" he asked.
>
> I called the census help line to find out and, sure enough, there was a button to press just for people with "questions about the meaning of race."
>
> "The concept of race reflects self-identification," a recorded voice said. "It does not indicate any clear-cut scientific definition which is biological or genetic in reference. The data for race represent self-classification by people according to the race or races with which they most closely identify." If that didn't make sense, try figuring out whether you are "Spanish/Hispanic/Latino" or just a plain old Chicano, Puerto Rican, or Cuban.
>
> At any rate, my white colleague would not go to jail for being black. As far as the Census Bureau is concerned, if a white person feels closely identified with blacks, so be it. He can be black for a day (or a decade, as the case may be).
>
> It did make me wonder though: How do we really know who's who out there? And does anybody really care? In 1995, the Post,

Harvard University and the Henry J. Kaiser Family Foundation conducted a survey in which most white people expressed the belief that blacks made up 23.8 percent of the U.S. population, nearly twice what the census says.

Maybe they were right. Maybe what they were saying is that they realize that there is no such thing as a "white" person, that we are all "colored" to one degree or another with blood from ancestors who can't be accounted for but which we all know have their origins in Africa.

The race category on the census form that really caught my eye was the one that supposedly applied to me. It came with three names attached: "Black, African Am., or Negro." I thought all of those were separate categories, with African Am. being some kind of airline.

African American, on the other hand, is the name most "people of color" prefer, according to recent opinion polls; black is no longer the in word. And speaking of the n-word, what about all of the black rappers who go by that? I can already smell an undercount.

As for "Negro," I hadn't seen one of them since 1968.

Race. What a mess.

Seeing all of the official racial distinctions based on a certain skin tone here and particular texture of hair there was to bear witness to a nation gone bonkers over a figment of its imagination.

Race, as we all know by now, is a biological fiction. It simply doesn't exist. Genetically, human beings are 99.9 percent the same. But we sure do make an awful lot of that 0.1 percent, mostly a cesspool of racism.

Last week, the U.S. Census Monitoring Board and the accounting firm of Price, Waterhouse, Coopers released a study estimating that certain metropolitan areas stand to lose $11 billion if the bureau repeats the undercount of 1990. African Americans were undercounted by about 4.4 percent, and Latinos were undercounted by 5 percent, the study noted.

A national campaign is now underway to get African Americans and Latinos to fill out the census forms. But getting an accurate count of people is one thing; counting by race is something else.

What is the point?

A 1992 poll by the Joint Center for Political and Economic Studies found that most Americans, including blacks and whites, have virtually the same concerns, hopes, and dreams. We all want to support our families, send our children to good schools and have adequate health care for the elderly.

Blacks are as likely as whites to invoke the virtues of individual responsibility, according to a Gallop poll, with more blacks than whites believing that black people must work harder to solve their problems and improve the lives of their families and themselves.

Earlier surveys by the Census Bureau found that blacks are the most cohesive group in the United States when it comes to reporting racial data. Only a handful of blacks report themselves as whites, compared with 18 percent of Latinos, the surveys show.

However, this race-based cohesion obscures some fundamental truths about our common humanity. And by emphasizing petty distinctions, we sometimes overlook similarities that could form the basis for powerful anti-racist coalitions.

One reason for the racial count in the census is supposedly to give the government a measuring stick to monitor civil rights violations, such as discriminatory lending practices by banks and mortgage companies. If we know how many blacks are living in an area, the theory goes, we can tell if they are being represented proportionally in politics, education and employment.

However, this leaves us with a most destructive paradox: By combating racism this way, we also give credence to the false concept of race, which is at racism's root.

And yet, not to acknowledge race is to allow the forces of racism to go unchecked.

What a mess.[1]

CONNECTIONS

A paradox consists of two true statements that seem to contradict one another. What is the paradox that Courtland Milloy sees in the 2000 census? Why does Milloy regard that paradox as "destructive"? To what extent do paradoxes like the one he describes foster illusions? Allow some to deny the reality of not only racism but also the diversity of the American people?

On the 2000 census, three of every 10 Americans described themselves as members of one of four minority groups—African Americans, Asian Americans, Native Americans, and Latinos. Approximately seven of every 10 Americans considered themselves white. The 2000 census provides no information on whether others view a given American as white, Latino, or something else. For the first time in American history, the 2000 census recognized the way individuals defined themselves as an important piece of information. If race is becoming a matter of "self-identification," what word or words describe the reality of racism—the negative ways some people view themselves and others based on skin color?

To what extent does the government's response to criticisms of its racial categories address the issues raised by critics? What are the implications of its response for individuals? For various groups within the nation? For the nation as a whole? Why do you think the government provided a button on its "census help line" just for people with "questions about the meaning of race"?

In the introduction to this chapter, Martha Minow noted that "when we simplify and sort, we focus on some traits rather than others, and we assign consequences to the presence and absence of the traits we make significant. We ask, 'What's the new baby?'—and we expect as an answer, boy or girl. That answer, for most of history, has spelled consequences for the roles and opportunities available to that individual." What traits does the census make significant? What consequences does it assign to the presence or absence of those traits?

For additional readings about identity and race, see Chapter 1 of *Facing History and Ourselves: Holocaust and Human Behavior.*

---

1. © 2000 *The Washington Post*, March 19, 2000, p. C01. Reprinted with permission.

"Passing"

Reading 5

For *The Twilight Zone*, Rod Serling created an episode which suggests beauty is "in the eye of the beholder" (Reading 2). Shirlee Taylor Haizlip believes that racial designations are also "in the eye of the beholder."

In *The Sweeter the Juice*, Haizlip describes her relatives by detailing her search for "lost" members of her mother's family. Haizlip's mother was separated from her closest relatives at the age of four. After the child's mother died, distant cousins reared her. She grew up thinking of herself as an African American. She also grew up wondering what had happened to her father, grandmother, uncle, aunts, sister, brothers, and cousins. They all "vanished" after her mother's death.

It took Shirlee Haizlip 15 years to locate her mother's missing relatives and learn their stories. Each had chosen to "pass" as white. In a magazine article in 1995, she reflects on the response to her book and what she learned from her research.

> The anthropologist Ashley Montagu was long an advocate of abolishing race as a concept. He never used the term except in quotation marks. Last year Dr. Luigi Cavalli-Sforza, a geneticist at Stanford University, confirmed that DNA is a potpourri of genes deriving from myriad ethnic sources. And Jonathan Beckwith, a microbiologist at Harvard Medical School, argues that scientists cannot measure genetic differences between the races.
>
> Yet "race," that socially constructed entity, was the reason for the breach in my mother's family. Although the two sisters had the same parents and skin color, one lived all her life as a black woman, and the other lived hers as a white woman, keeping her black heritage a secret from her white husband, their only child, and their grandchildren. The sister was not alone in the choices she made. My mother's other siblings and the rest of her family had also abandoned their race. They acted on the complexly simple infinitive "to be," and in fact they "were," and their descendants still "are" . . . "white."
>
> Some would say these relatives have "one drop" of black blood, so they are in fact black. But except in Louisiana all of the "one drop" racial laws have been rescinded since 1986. So if you look white, marry white, live in a white community, attend a white church and a white school, join white associations, have white looking children and grandchildren, you are "white," as defined by the majority in this country.

Hundreds of thousands of blacks passed for white, starting in the days of slavery and continuing into the present. Because of the secret nature of the transaction, no records were kept of the exact numbers who created new places for themselves in American society. Population experts tell us that large numbers of black people are "missing." I doubt that they were abducted by aliens.

According to Carla K. Bradshaw, a clinical psychologist and professor at the University of Washington, "Passing is the word used to describe an attempt to achieve acceptability by claiming membership in some desired groups while denying other racial elements in oneself thought to be undesirable. The concept of passing uses the imagery of camouflage, of concealing true identity or group membership and gaining false access. Concealment of 'true' identity is considered synonymous with compromised integrity and impostorship. . . . If an ideal world existed free from the psychology of dominance, where racial differences carried no stigma and racial purity was irrelevant, the concept of passing would have no meaning. In fact, passing of any kind loses meaning in the context of true egalitarianism. . . . "

Some geneticists claim that as many as 80 percent of black Americans have white bloodlines and that a surprising 95 percent of white Americans have some black ancestry. These statistics are based not on guesswork but on the direct clinical examination of nucleotides and microsatellites, genetic components common to all human blood. Dr. Luigi Cavalli-Sforza tells us . . . that modern Europeans (the ancestors of America's immigrants) have long been a mixed population whose genetic ancestry is 65 percent Asian and 35 percent African. There has never been any such thing as a "Caucasoid" gene, nor is there such a creature as a "pure" white or black American. . . .

Just from looking at archival records of my family, I know that every census has measured race differently. In different periods the same people in my family were listed as mulatto, black, or white. The designation could depend on the eye of the beholder or the neighborhood where they lived. In the meantime, their neighbors, their co-workers, and their communities at large saw them as either black or white, depending on who decided what.[1]

CONNECTIONS

How have members of Shirlee Haizlip's family answered the question: What do you do with a difference? What are the consequences of their responses?

What is "passing"? How do myths and misinformation about "race" explain the practice? How does the fact that thousands of "blacks" have successfully "passed" as "whites" reinforce the idea that "race" is a "social invention" rather than a scientific description of human differences? Shirlee Haizlip believes that "passing" is a way of coping with the legacy of slavery. To what extent may "passing" also be a mechanism for survival during the years of "Jim Crow laws"—laws that isolated and humiliated African Americans?

Some anthropologists believe that the first step in eliminating racism is separating our need to belong from the dangerous temptation to hate others. What is the connection between the two? How would you go about separating them? The film *Twilight: Los Angeles, 1992* reveals some of the difficulties in resolving racial conflicts. Copies of the film and a study guide prepared by Facing History and Ourselves are available from Facing History and Ourselves resource centers.

In her novel *Paradise,* Toni Morrison meditates on questions of difference. A number of readers have noticed that she never mentions the "race" of several women in the book. When asked why, Morrison said she wanted "to have the reader believe—finally—after you know everything about these women, their interior lives, their past, their behavior—that the one piece of information you don't know, which is, the race, may not, in fact matter. And when you do know it, what do you know?" How would you answer her question?

What does Shirlee Haizlip mean when she claims that racial designations are in "the eye of the beholder." How have classifications changed from one census to the next? Find out what prompted those changes.

Two videotaped interviews, available from Facing History and Ourselves, provide insights into the impact of racism and antisemitism on identity in the mid-20th century. *Facing Evil* features author Maya Angelou reading her adaptation of Paul Lawrence Dunbar's poem, "We Wear the Mask," to reveal the hidden pain that she and many other African Americans experience daily during the years of segregation. *Childhood Memories* features sociologist Nechma Tec's account of her childhood in Nazi-occupied Poland. Passing for her was a life-or-death matter. She had to hide her Jewish identity to avoid the death camps.

---

1. "Passing" by Shirlee Taylor Haizlip. Reprinted by permission of AMERICAN HERITAGE magazine, a division of Forbes, Inc., © January/February, 1995, pp. 47-49.

Reading 6

If race is a "social invention," a "biological fiction" as Shirlee Haizlip believes, what is racism? When asked to explain the term, Lisa Delpit—a scholar, a teacher, and the author of an influential book about race and education entitled *Other People's Children*—expressed her ideas in the form of a letter to her nine-year-old daughter. It says in part:

> My Dearest Maya,
>
> You are amazing. Your golden brown skin, your deep black "ackee" eyes, your wiry, gold-flecked hair that seems persistently unwilling to stay contained in any manner of braid or twist I devise. I listen in amazement at your interpretations of the world and laugh at your corny nine-year-old's jokes. I can't fathom how you've managed to turn those little baby digits I loved to kiss into the long, graceful fingers—adorned at the tips in blue and purple designer colors—that now dance so expertly across your violin strings. Yes, you are amazing.
>
> As much as I think of you as my gift to the world, I am constantly made aware that there are those who see you otherwise.
>
> Although you don't realize it yet, it is solely because of your color that the police officers in our predominantly white neighborhood stop you to "talk" when you walk our dog. You think they're being friendly, but when you tell me that one of their first questions is always, "Do you live around here?" I know that they question your right to be here, that somehow your being here threatens their sense of security.
>
> I didn't tell you exactly what was going on when we took that trip to the Georgia mountains. You and your friend played outside the restaurant while his mom and I visited the ladies room. Later, the two of you told us that a white man and his wife—he with a minister's collar—stared at you "with mean looks" and made monkey sounds and gestures. You asked why they did that, and I told you that some people were just not nice. I made you promise to come to me immediately whenever an adult was giving you trouble.
>
> I did not have to be told much when I was your age. When I was growing up in Louisiana in the 1950s and 1960s, the color lines were very clearly drawn. I followed my mother to the back entrance of the doctor's office, marked "colored." I knew which

water fountain I was supposed to drink from. On the bus ride to my all-black school, I watched white children walk to schools just two or three blocks from my house.

In large part, my childhood years were wrapped in the warm cocoon of family and community who all knew each other and looked out for one another. However, I remember clearly my racing heart, my sweaty-palmed fear of the white policemen who entered my father's small restaurant one night and hit him with nightsticks, the helpless terror when there were rumors in our school yard that the Ku Klux Klan would be riding, the anxiety of knowing my college-aged foster sister had joined the civil-rights marchers in a face-off against the white policemen and their dogs. And, I remember, my Maya, the death of your grandfather when I was seven, who died of kidney failure because the "colored" ward wasn't yet allowed the use of the brand-new dialysis machine.

Your world is very different, at least on its surface. In many ways now is a more confusing time to live. In *Seeing a Colorblind Future*, Patricia Williams says we are saturated with insistent emblems of brotherhood—multicolored children singing "We Are the World;" television shows with the obligatory child of color; teachers' adamant statements that "we are all the same" and "color doesn't matter." Yet, attacks on rectifying past discrimination are made unabashedly under the flag of "color-blindness," white hate crimes are on the upswing, many communities and schools are more segregated than they were 20 years ago. I receive at least a call a week from frantic African-American parents living all over the country who are terrified at the hostility shown regularly by the schools to their brown children.

As any mother would, I have a great need to protect you, but it is hard to know how. My childhood experience was different from yours. As was the case in many African-American Louisiana families, our family was a rainbow of colors from chocolate-brown brunettes to peach-colored blondes. (The history of that reality is a story we'll need to talk about later.) I was the light-skinned, freckled, red-headed child, who always got the sunburn whenever we went to the beach. Because of my coloring, I had another role, too. When traveling by car, African Americans were not allowed to use the restrooms or other facilities white travelers took for granted. Black families had to develop all sorts of strategies to make a road trip workable. When it was time for a rest stop, one of our ruses was to pull around to the side of the service station and send in the one who looked most like white to get the key. Then, outside of the attendant's view, everyone would use the facility.

Decades later, when you were an infant, your aunt and I drove to Mississippi. I had not made that trip for many years, and although segregation was officially over, I still felt uneasy at the rest stops. Any African American would. There were Confederate flags printed on every possible souvenir in the gift shops, and restaurants and gas stations were filled with burly, white, cigarette-smoking men with gun racks mounted in their rear windows. Heart racing, cradling my beautiful brown baby, I suddenly realized I did not know how to protect you from the vicious hatred in some of the eyes that stared at us. Or, for that matter, from a society whose very structure privileges some and marginalizes you.

I have tried to protect you from the disease of internalized racism—of seeing yourself through the eyes of those who disdain you—that infects the souls of so many of our young people. When I was in my segregated, all-black elementary school, we were told by teachers and parents that we had to excel, that we had to "do better than" any white kids because the world was already on their side. When your cousin Joey was in high school, I remember berating him for getting a "D" in chemistry. His response was, "What do you expect of me, the white kids get "C's." Recently a colleague tried to help an African-American middle-schooler to learn multiplication. The student looked up at the teacher and said, "Why are you trying to teach me this? Black people don't multiply. Multiplication is for white people." You know, Maya, I think that may be the biggest challenge you and other brown children will face—not believing the limits that others place upon you.

It is not easy to know how to keep you believing in yourself, even believing in your abundant radiance and beauty. I know there was a time when you couldn't understand why I wouldn't allow you to wear a white character mask at Halloween, or why I told your grandmother to stop sending you white dolls.

It's hard for a mother to know just how far to go with principles, though. And I think you helped me develop a somewhat less strident attitude in your own brilliant, unpredictable way. I remember refusing to buy a white Barbie—even though the store didn't have the black one with equivalent turn-colors-in-the-sun hair. You were not happy with me, even though I explained at length the reasons regarding not bringing dolls into our family who looked like they would not possibly be a part of our family. "You don't see any of your white friends begging for a black Barbie doll, do you?" I asked, adding what I thought would be the final word. But several days later in another conversation, you asked, "Mom, do you have any white friends?" "Of course, I do, Maya, you know that," I answered. "Do you like your white friends, Mom?" "What a question, Maya, if they're my friends, then I like them." "Well, Mom," you delivered your knock-out punch, "my black Barbies want some white friends, too." Well, my dear, from that moment on your doll collection became interracial.

It is so hard to know how both to engender the possibility of color not mattering—where people will truly be judged not by the color of their skin, but by the content of their character—and to give you understanding that will create a protective armor for the real world of racial bias that exists around you. I don't want to limit you, to have you always on edge (as I sometimes feel) questioning the intentions of white playmates or teachers. Decisions based on color are so pervasive, and people of color so demonized in this country—though racist comments are often thinly camouflaged by such terms as "teen-aged mother," "the criminal element," "welfare cheaters," "drug dealers," "school drop-outs," "at-risk students"—that understanding societal realities does not come as easily as it did in my childhood.

Yes, Maya, I really do want to believe that a color-blind future is possible. I never express my doubts when one of the parents at your school calls at the last minute to invite you to a birthday party, adding that "Suzie [whom you hardly know and seldom play with, but who is the only other black girl in your class] is coming."

I am proud yet torn when I hear you come to some understanding on your own. Like when you were seven and playing with a little friend who had brought his cowboy and Indian figures, and you said, "OK, I'll be the Indians and you be the bad guys." Or when you went bike-riding with a friend and came back upset that "a white boy"—as opposed to just "a boy"—said he was going to hurt you. Or when you asked me why there weren't any black teachers in your school and added that you hoped that the school "didn't think black people weren't as smart as white people." When I told you that you needed to talk to the principal about that, you went right up to her the next day and asked your question. She, to her credit, took your question seriously and explained that they would like to find more black teachers, but that the salaries the school paid made it hard to attract them. Not one to let anyone off easy, you immediately came back with, "Well have you tried Morehouse?"

I am pleased that you have realized that brown skin is good. Yet I am saddened that you cannot be innocent to the unfortunate realities surrounding you. You have understood that the color line lives.[1]

CONNECTIONS

What adjectives does Lisa Delpit use to describe the racism she experienced as a child? How did it shape her attitudes toward white Americans? What changes have taken place since her childhood? How have those changes affected the way she sees herself and others? The way she is rearing her daughter?

Create an identity chart for Lisa Delpit as a young girl. Begin with the words or phrases that Delpit uses to describe herself. Then add the labels others might attach to her. Create similar charts for Delpit as an adult and for her daughter. What words or phrases does Delpit use to describe Maya? What words or phrases might Maya use to describe herself? What words or phrases might Maya use to describe her mother?

How are the three charts alike? What differences seem most striking? What part has race played in shaping the identity of both mother and daughter? What part has racism played in shaping their identity? Compare these charts to your own identity chart. What do you have in common with the Delpits? What part has race played in shaping your identity? What part has racism played?

Write a working definition of the word *race*. A working definition is one that grows as you read, reflect, and discuss ideas. Begin your definition by explaining what the word *race* means to you. Then add the meanings implied in each of the readings you have completed in this chapter. Next create a working definition of the word *racism*. Keep in mind that the ending *ism* refers to a doctrine or principle. Can you be a racist if you do not believe in the concept of "race"? Expand your definitions as you continue reading.

How does Delpit define the term *internalized racism*? Why does she call it a "disease"? Based on your own reading and experiences and those of your friends, how would you define the term? What would you add to her definition? What would you change?

Why do you think Delpit tells her daughter that "in many ways now is a more confusing time to live"? In what sense is it more confusing today? In her view, how does that confusion shape the way the young African Americans she knows view their identity?

Although Lisa Delpit's letter is addressed to her daughter, it was written for a wider audience. She later told an interviewer that she wanted to capture the "torment that I, her mother, face each time I am confronted with racism's ugly face." Why do you think she chose to voice her views in such a personal way? What feelings and emotions would have been more difficult to express in an essay that takes a more scholarly approach to the question? Write an essay explaining your views of racism. If you choose to write your essay as a letter, think about to whom it should be addressed—your parents, a teacher, a younger brother or sister, or perhaps Lisa Delpit and her daughter Maya. Keep a copy of your essay in your journal or a portfolio so that you can revise, expand, or rewrite portions of it as you continue reading this book.

---

1. "Explaining Racism" by Lisa Delpit in *The Magazine of the Harvard Graduate School of Education*, Spring, 2000, pp. 15–17.

Reading 7

Lisa Delpit doesn't want labels "to limit" her daughter's potential. Nancy and Stanton Wolfe have similar concerns about their daughter. They too worry about the power of labels. Their daughter Ashley has Down syndrome—that is, she was born with an extra chromosome 21. A child with Down syndrome is usually smaller and his or her physical and mental development is slower than a youngster who does not have Down syndrome.

Ashley sees herself as more than the sum of the various labels people attach to individuals with Down syndrome. She told reporter Bella English, "My goal is to change the way people think about us. I do have special needs, but I have special abilities. I just want to be seen as who I am." In an interview, English discovers who Ashley is. She writes in part:

> She received a standing ovation when she spoke at her high school graduation. She's currently in her third year at Lesley College in Cambridge. She recently played a role in the NBC drama, "Third Watch." Two afternoons a week, she volunteers at Massachusetts Eye and Ear Infirmary. She has a new job at Harvard's Fogg Art Museum. She also has Down syndrome.
>
> When Ashley Wolfe was born 21 years ago, her parents knew at once something was wrong, though "wrong" is not the word they would choose. Nancy and Stanton Wolfe consider their daughter pretty perfect the way she is. "She just has a little extra chromosome," her mother says.
>
> And that is how Ashley Wolfe has tried to live her life: putting that extra chromosome in its place. "Having Down syndrome is just one little part of me," she says. On a recent day, she looks pretty much like any other young person: wearing jeans, a red shirt, gold star earrings, pouring a cup of tea for a visitor in the Union Square apartment she shares with two other students.
>
> Yes, there are the vision problems, the speech that will slur if her vigilant attention to enunciation drifts, the gait problems that come from having one leg shorter than the other, the social problems that dog those with Down syndrome. But she wants people to know there's more to her than a medical diagnosis. "Back in the early ages," she says, "people with Downs were called mongoloids and they would institutionalize them. My parents really wanted me to be mainstreamed. I'm glad."

It was in the recovery room that the Wolfes learned for certain their newborn had Down syndrome, a genetic condition that causes multiple problems, including mental retardation, and occurs in one of every 800-1,000 live births.

The pediatrician arrived just after Ashley and said two things the Wolfes will never forget: "She's beautiful," and "Her heart's perfect." Many Down babies have serious heart defects. For Ashley's good heart, her parents were grateful.

But there were other folks whose comments cut deeply. "Are you taking her home?" was a question the parents heard often those first several hours.

There was never any doubt that Ashley would be well loved and supported. From the start, the Wolfes wanted her to be mainstreamed, which meant that she was in many regular classes, where she had her own aide. For other subjects, she went to the resource room, which offered smaller classes for slower learners. By the time she graduated from Simsbury High School outside Hartford, she had made the honor roll seven times. She had even taken Spanish.

"I had to work extremely hard to do that," says Ashley. "My parents made me study extremely hard.". . .

But her crowning achievement thus far came on graduation night in June 1997. She stood at the podium before 3,000 people and delivered a powerful message of hope called "Opening Eyes, Opening Minds." She and several other students had auditioned in front of a faculty committee; only three were chosen. "I wrote it," says Ashley, "but I had help from my parents and speech therapist."

Principal Dennis Carrithers remembers the speech well. "It was one of the most beautiful things I've ever seen at any school," he says. "She spoke about the things she learned here, the people who mattered to her. When she finished, people were on their feet, wiping away tears." "She's a really strong lesson that we never want to set limits on people," Carrithers says, "because we have these wonderful surprises like Ashley."

That's not to say life has been easy. "It's been a very big struggle for both me and my parents," Ashley says, sitting at her kitchen table, her appointment book in front of her. She is proud of the fact that she is organized, right down to a list of questions to ask the reporter. ("When will the story run? What section?") "I have to write everything down," she says, "or I might not remember."

The social issues have been as painful as the physical ones. "People said, 'Does she have potential? Is there a future for her?'"

she recalls. "In school, labels are put on. 'Oh, you're a special ed student.' The normal kids didn't want to be around us. I had very few friends."

It's not that other children overtly teased Ashley, her mother says, but neither did they seek her out. "We joined the Brownies and stuff like that," Nancy Wolfe says, "but I think she was always unsure of herself around kids her own age."

When Ashley was 7, she underwent several operations for dislocated hips and spent two years in a cast from her toes to her waist. "I basically had to learn to walk again," she says. Years of physical and speech therapy followed.

When complimented on her speech, she smiles. "Thank you," she says. "It took a long time. Every once in a while, I do get lazy with my speech. I'll have people tell me to please repeat things."

Obviously, Ashley Wolfe is on the high end of those with Down syndrome. Her main cognitive problems are with math and directions, and health issues remain. She is not able to drive.

To help her with time—she has trouble with clock faces—her parents bought her a digital watch. "She's always way early, just to protect herself," says Nancy Wolfe, an actress who runs a summer arts program at Wesleyan University. Money is another problem: she simply has difficulty handling it.

Ashley describes her limitations this way: "I have a very hard time with integration. That basically means putting things together, like walking into a situation and making sense of it."

One of her mother's favorite pictures is of Ashley as a 3-year-old. "She had these long blond pigtails, an eye patch, glasses and braces on her legs, and she was dancing around the living room," says Nancy Wolfe. '"She has this irrepressible spirit. Sometimes, she calls me up and cries. I just say, 'No, it isn't fair, Ash.' If you had told me three years ago that she would be living on her own and balancing her checkbook I wouldn't have believed it. She has continued to raise the bar for us."

Her father describes her as "nothing short of a miracle." His expectations for her? "I never allowed myself expectations," he says, "but I also never had limitations, and I think that is key. I felt there was no limit on what she could accomplish." Her greatest achievement? "Who she is."

That Ashley has achieved so much is due in large part to her family. Stan Wolfe is a facial surgeon who recently went back to school and earned a master's degree in public health. He is now oral health director, as well as supervisor of school and primary health, for the Connecticut Department of Public Health. Nancy Wolfe has worked with multiply-handicapped kids in the theater. The couple were determined to give Ashley the most normal life they could.

But perhaps their greatest gift was Rebecca. '"Rebecca," says Ashley with a smile, "is wonderful."

Rebecca is the sister who arrived 3½ years before Ashley. A magna cum laude graduate of Harvard, Rebecca also lives in Somerville. One of her earliest memories is being told that her very special sister had just been born. One of her best memories is Ashley's graduation speech. "It's one of those things," she says, "where you felt bad for the person who had to go after her."

"We are very close,'' Ashley says. "I take the bus to her house."

"Shley" is what Rebecca calls her younger sister. She is unabashedly proud of her, and has always felt more a little mother than a big sister. "There's a lot of sadness for me around not ever having a normal sister relationship," she concedes. When she was in college, Rebecca Wolfe wrote a story for a student magazine called, "Hero Worship: How Down Syndrome Challenged the Love Between Two Sisters." In it, she recounted the fierceness with which she protected her sister—and the embarrassment she sometimes felt.

"I'd spent my elementary years terrified that someone would make fun of me for Ashley. I hated myself for feeling even a little ashamed of her, and dared them to try it,'' she wrote. ''I will always have conflicting and confusing emotions of love, admiration, frustration, and sadness for her."[1]

CONNECTIONS

Create an identity chart for Ashley Wolfe. How is it like the ones you created for Lisa Delpit and her daughter Maya (Reading 6)? How is it like your own identity chart? What differences seem most striking?

How have Ashley Wolfe's parents and sister answered the question: What do you do with a difference? How do you think their expectations, attitudes, and beliefs have shaped Ashley's identity?

At Harvard, Rebecca Wolfe compiled a book of photographs from Ashley's life and helped her sister put words to the pictures. She called the book *A Different*

Kind of Knowledge. What do you think the title means? How does it relate to Ashley's life? In the book, Ashley writes: "I am not too sure about the future. I just do the best I can do. Maybe have a job, get married, have kids. Because everyone has to work because we need money and we need to find love and hope." How does she seem to answer the question: What do you do with a difference?

How does Ashley Wolfe's story complicate our understanding of the meanings we attach to difference? What does her story suggest about how our choices make a real difference in the world?

What kind of education would help young people better understand the meanings we attach to differences? Would encourage them to challenge the social consequences of particular difference?

---

1. "An Everyday Courage Despite Battles Physical and Emotional, 21-year-old Ashley Wolfe Remains Irrepresible" by Bella English. *The Boston Globe,* March 9, 2000. © Copyright 2000 Globe Newspaper Company.

Reading 8

Like *The Twilight Zone* in the 1960s (Reading 2), a TV series in the 1990s entitled *Star Trek: The Next Generation* often dealt with the question: What do you do with a difference? In one episode, the crew of the starship *Enterprise* encounters a society that uses genetic engineering to eliminate differences. The episode entitled "The Masterpiece Society" is available at many video stores or may be borrowed from Facing History and Ourselves. If possible, watch the episode as a class. If you are unable to obtain the video, the following paragraphs summarize the story.

> The starship *Enterprise*, representing the United Federation of Planets, enters the Moab star system on a research mission. The crew is tracking the course of a stellar core fragment—a massive remnant of a supernova—that is passing through the apparently uninhabited Moab system.
>
> As the fragment draws closer to the desolate planet Moab IV, the *Enterprise* science officer discovers that an artificial environment on the planet shelters a human society. The entire population is in danger. The stellar fragment will trigger huge earthquakes that will destroy everyone.
>
> Captain Picard explains to the planet's chief administrator, Aaron Connor, why it may be necessary to evacuate the planet. In the face of the emergency, Connor reluctantly agrees to allow several members of the *Enterprise* crew to visit Moab IV and search for ways to save the planet.
>
> Connor explains to the crew that evacuation is impossible. "You see, this is an engineered society . . . genetically engineered. Our ancestors came from Earth to create a perfect society. They believed that through controlled procreation they could create people without flaws, and those people would build a paradise."
>
> Connor's adviser, Martin, elaborates, "We have extended the potential of humanity, physically, psychologically. We have evolved beyond . . . beyond—"
>
> "Beyond us," chief engineer Geordi La Forge of the *Enterprise* remarks acidly.
>
> "Frankly, yes," Martin agrees. "No one in this society would be blind, for example," he adds, looking pointedly at the vision-visor that covers La Forge's sightless eyes.

Tactfully, Connor seeks to smooth things over. He explains that all the living things on the planet are interconnected: "We are part of our environment, and it is part of us. . . . Each of us grows up knowing exactly what our society needs from us. . . . Let me put it this way. Are there still people in your society who have not yet discovered who they really are or what they were meant to do with their lives? They may be in the wrong job. They may be writing bad poetry. Or worse yet, they may be great poets working as laborers, never to be discovered. That does not happen here!"

When Counselor Deanna Troi reports that the people of Moab IV may risk death rather than give up their way of life, Captain Picard is appalled. "They've given away their humanity with this genetic manipulation. Many of the qualities that they breed out—the uncertainty, the self-discovery, the unknown—are many of the things that make life worth living, at least to me. I wouldn't want to live my life knowing that my future was written, that my boundaries had been already set, would you?"

Troi answers, "I don't know. I doubt it. Nevertheless, it is what they believe in."

A theoretical physicist from the planet, Hannah Bates, comes aboard the *Enterprise* to work with La Forge to find a way to shift the stellar fragment's course. They continue to discuss the problems of genetic engineering. "Were you always blind?" Hannah asks.

"I was born blind," Geordi says. "I guess if I had been conceived on your world, I wouldn't even be here now, would I? I'd have been terminated as a fertilized cell."

As Hannah and Geordi work on the problem, their solution is inspired by the technology used in Geordi's visor. "That's perfect!" he exclaims. "The answer for all of this is in the visor created for a blind man who would never have existed in your society."

While one crisis is averted, another has been developing. A number of the people from Moab IV now wish to leave the planet to learn more about life beyond its boundaries. But because the society is genetically integrated, such departures will create gaps. "If even a handful leave," Connor tells Picard, "the damage to this society will be devastating. . . . Thousands will suffer."

The *Enterprise* crew debates whether to offer asylum to those who wish to leave. Ultimately, about two dozen people from the planet leave with the starship. "In the end," Picard muses, "we may have proved just as deadly to that society as any core fragment."[1]

CONNECTIONS

How do the people of Moab IV answer the question: What do you do with a difference? What arguments do their leaders use to defend genetic engineering? How do the crew and the captain of the *Enterprise* respond to those arguments?

What does Captain Picard mean when he says the people of Moab IV have "given away their humanity"? How does he seem to define *humanity*? How do the people of Moab seem to define the word? How do you define it? How are the definitions similar? How do you account for differences?

The word *irony* describes a contrast between what is stated and what is meant or between what is expected to happen and what actually takes place. What are the ironies in the solution to the crisis that the people of Moab IV face? What do those ironies reveal about the way the creators of *Star Trek: The Next Generation* view a "masterpiece society"? How do you regard such a society?

In 2000, scientists announced the completion of the first survey of the entire human genome. The survey is helping researchers eliminate birth defects and other health problems. In time, it may also help them identify certain "beneficial" genes that detoxify the body and resist disease. According to *The Scientist*, researchers are now able to "examine the role of noncoding elements of the genome—introns—whose differing sizes across species, it turns out, may play a direct role in modulating gene expression levels, even shaping individual human differences of thought, morphology, and personality." Find out more about the various research projects that are part of the Human Genome Project and share your findings with the class. What questions does the research raise about what it means to be human?

"Harrison Bergeron," a short story by Kurt Vonnegut, imagines a society in which differences have been outlawed so that everyone is truly equal. No one is superior to anyone else. The story is reprinted in Chapter 2 of *Facing History and Ourselves: Holocaust and Human Behavior.* Is the society Vonnegut describes fair? Is it just? How is it like Moab IV? How do you account for differences?

---

1. Adapted from "The Masterpiece Society" (Episode 113) *Star Trek: The Next Generation.* Copyright ©1992 by Paramount Pictures.

2. Race, Democracy, and Citizenship

We hold these truths to be self-evident: That all men[1] are created equal; that they are endowed by their Creator with certain inalienable rights; that among these are life, liberty, and the pursuit of happiness.

Declaration of Independence

Chapter 1 explored the physical characteristics, social ties, and other factors that shape our identity. The chapter also considered the consequences of the ways we view the differences between ourselves and others. The readings in Chapter 2 place those ideas in historical perspective by examining how Europeans and Americans regarded differences in the 1700s and early 1800s. Many of the beliefs that Americans hold today about race and democracy developed during those years. It was a time when hundreds of thousands of Africans were brought to the Americas in chains. Their enslavement had a profound effect on American attitudes and values then and now. The 1700s and early 1800s were also a time when some Europeans and Americans participated in an intellectual movement known as the "Enlightenment." In 1784, Immanuel Kant described the "enlightened" as those who "dare to know, to reject the authority of tradition, and to think and inquire for oneself." Modern science grew out of the ideals of the Enlightenment. So did many democratic institutions.

Among those attracted to the Enlightenment were the leaders of the American Revolution. Indeed, the movement inspired many of the ideas central to citizenship in the United States, including the belief "that all men are created equal." Yet it was also the Enlightenment that encouraged the notion that humankind is divided into distinct and unequal races. It was an idea supported by scientists who exaggerated the differences between *us* and *them* to justify prejudice, discrimination, and slavery.

Many of the readings in this chapter consider not only the tension between these two contradictory notions about human worth—racism and equality—but also the consequences of that tension and its effects on the lives of real people long ago and today. In analyzing their stories, it is important to remember that the thinkers of the Enlightenment lived at a particular time and in a particular place. The great events, prejudices, and values of that time and place shaped thinking, much as they shape thought today. Anne Fausto-Sterling, a professor of biology and medicine, says of the relationship between science and society:

> Scientists peer through the prism of everyday culture, using the colors so separated to highlight their questions, design their experiments, and interpret their results. More often than not their hidden agendas, non-conscious and thus unarticulated, bear strong

> resemblances to broader social agendas. Historians of science have become increasingly aware that in even the most "objective" of fields—chemistry and physics—a scientist may fail to see something that is right under his or her nose because currently accepted theory cannot account for the observation.[2]

Aware that their work reflects the values of their society, many scientists today have come to believe that scientific research is more than a method of inquiry. It also requires a willingness to challenge dogma and a willingness to see the universe as it really is. Accordingly, science sometimes requires courage—at the very least, the courage to question the conventional wisdom. British scientist P. B. Medawar once wrote of his own research:

> I have been engaged in scientific research for about fifty years and I rate it highly scientific even though very many of my hypotheses have turned out mistaken or incomplete. This is our common lot. It is a layman's illusion that in science we caper from pinnacle to pinnacle of achievement and that we exercise a Method which preserves us from error. Indeed we do not; our way of going about things takes it for granted that we guess less often right than wrong, but at the same time ensures that we need not persist in error if we earnestly and honestly endeavor not to do so.[3]

Chapter 2 considers what happens to a society when some leading scientists are unwilling or unable to accept the idea that their hypotheses could possibly be "mistaken or incomplete." It also explores how the "twisted science" that results from such research becomes the "conventional wisdom"—the things we are so convinced are true that they are rarely if ever challenged.

---

1. The words *men* and *mankind* were commonly used in earlier centuries to refer to humans and humankind. Their use reflects a particular time period.
2. *Myths of Gender* by Anne Fausto-Sterling. Revised edition. Basic Books, 1992, pp. 9-10.
3. *The Limits of Science* by P. B. Medawar. Harper & Row, 1984, p. 101.

Reading 1

In the mid-1700s, a few European thinkers tried to apply the ideas and methods of science to humans and human societies. These thinkers were part of a movement known as the "Enlightenment." Although they disagreed on a number of points, most came to believe that all humans everywhere have the ability to reason and form societies. In time, those theories shaped the way ordinary people viewed the world. If societies are human inventions, some argued, people may alter or even replace an oppressive government with one more to their liking. In 1776, thirteen colonies along the eastern coast of North America broke their ties to Britain. Thomas Jefferson of Virginia wrote their Declaration of Independence. It states:

> We hold these truths to be self-evident: That all men are created equal; that they are endowed by their Creator with certain inalienable rights; that among these are life, liberty, and the pursuit of happiness. That, to secure these rights, governments are instituted among men, deriving their just powers from the consent of the governed; that, whenever any form of government becomes destructive of these ends, it is the right of the people to alter or abolish it, and to institute a new government, laying its foundation on such principles, and organizing its powers in such form, as to them shall seem most likely to effect their safety and happiness.

The French Revolution was also inspired by the ideals of the Enlightenment. In 1789, in their Declaration of the Rights of Man and Citizen, the French boldly stated that "Men are born, and always continue, free and equal in respect of their rights." As a French leader explained, "Since men are all made of the same clay, there should be no distinction or superiority among them." He and other thinkers of the Enlightenment regarded human differences as differences in degree rather than in kind.

Although the great thinkers of the Enlightenment stressed the equality of humankind, the notion that humanity is divided into separate but unequal races developed during those same years. Historian Londa Schiebinger offers one explanation:

> The expansive mood of the Enlightenment—the feeling that all men are by nature equal—gave middle- and lower-class men, women, Jews, Africans, and West Indians living in Europe reason to believe that they, too, might begin to share the privileges heretofore reserved

for elite European men. Optimism rested in part on the ambiguities inherent in the word "man" as used in revolutionary documents of the period. The 1789 [French] Declaration of the Rights of Man and Citizen said nothing about race or sex, leading many to assume that the liberties it proclaimed would hold universally. The future president of the French National Assembly, Honoré-Gabriel Riqueti, comte de Mirabeau, declared that no one could claim that "white men are born and remain free, black men are born and remain slaves." Nor did the universal and celebrated "man" seem to exclude women. Addressing the Convention in 1793, an anonymous woman declared: "Citizen legislators, you have given men a constitution . . . as the constitution is based on the rights of man, we now demand the full exercise of those rights for ourselves."

Within this revolutionary republican framework, an appeal to natural rights could be countered only by proof of natural inequalities. The marquis de Condorcet wrote, for instance, that if women were to be excluded . . . one must demonstrate a "natural difference" between men and women to legitimate that exclusion. In other words, if social inequalities were to be justified within the framework of Enlightenment thought, scientific evidence would have to show that human nature is not uniform, but differs according to age, race, and sex.

Scientific communities responded to this challenge with intense scrutiny of human bodies, generating countless examples of radical misreadings of the human body that scholars have described as scientific racism and scientific sexism. These two movements shared many key features. Both regarded women and non-European men as deviations from the European male norm. Both deployed new methods to measure and discuss difference. Both sought natural foundations to justify social inequalities between the sexes and races.[1]

CONNECTIONS

What does the idea of equality mean to you? Does the idea that "all men are created equal" imply that there are no differences? To what extent have Americans achieved equality? To what extent does it remain a dream? What is the power of that dream? How does that dream relate to citizenship?

Londa Schiebinger regards the development of scientific racism and sexism at a time when the great thinkers of the Enlightenment were stressing the equality of humankind as a paradox. How does she account for contradictions? How do you account for them?

According to Schiebinger, the thinkers of the Enlightenment encouraged a search for "evidence . . . that human nature is not uniform, but differs according to age, race, and sex." What does she suggest inspired that search? What consequences might such a search have? Suppose the Enlightenment had sought evidence of the similarities among humankind. What might have been the consequences of that search for society? For individuals within that society?

---

1. *Nature's Body: Gender in the Making of Science* by Londa Schiebinger. Copyright ©1993 by Londa Schiebinger. Reprinted by permission of Beacon Press, Boston, pp. 143-145.

Reading 2

The previous reading focused on how the great thinkers of the Enlightenment viewed equality and difference. Their ideas helped shape the way ordinary people viewed the world. Ideas are also shaped by the experiences of everyday life. Jack Foley traces the development of the notion that Europeans are white to the growth of slavery in the British colonies:

> According to the *Oxford English Dictionary*, the first appearance in print of the word white meaning "a white man, a person of a race distinguished by a light complexion" was in 1671. The second was in 1726: "There may be about 20,000 Whites (or I should say Portuguese, for they are none of the whitest) and about treble that number of Slaves." The term *Caucasian* is even later: "Of or belonging to the region of the Caucasus; a name given by [Johann] Blumenbach (ca. 1800) to the 'white' race of mankind, which he derived from this region."
>
> "Through the centuries of the slave trade," writes Earl Conrad, in his interesting book, *The Invention of the Negro*[1], "the word race was rarely if ever used. . . . [William] Shakespeare's Shylock uses the word *tribe, nation,* but not *race.* The Moor in Othello calls himself black and the word slave is several times used, but not race. The word does not appear in the King James Version of the Bible in any context other than as running a race. The Bible refers to nations and says: 'God made the world and all things therein; and hath made of one blood all nations of men for to dwell on all the face of the earth.' The Bible, with all its violence and its incessant warfare between peoples, does not have racist references to tribes, groups, provinces, nations, or men."
>
> And again, on the subject of slavery: "The traffic grew with the profits—the shuttle service importing human chattel to America in overcrowded ships. It was on these ships that we find the beginnings—the first crystallizations—of the curious doctrine which was to be called 'white supremacy.' Among the first white men to develop attitudes of supremacy were the slave ship crews."[2]

Colonial charters and other official documents written in the 1600s and early 1700s rarely refer to British colonists as white. By the late 1700s, however, the word was widely used in public documents and private papers. According to scholar Leon Higginbotham, Jr., it was also becoming entwined with the idea of citizenship. Increasingly, states viewed a citizen as a man who could help his neighbors put down slave rebellions or fight a war against the Indians. That notion of citizenship was reflected in the Naturalization Act of 1790. It states:

> All free white persons who, have, or shall, migrate into the United States, and shall give satisfactory proof, before a magistrate, by oath, that they intend to reside therein, and shall take an oath of allegiance, and shall have resided in the United States for one whole year, shall be entitled to the rights of citizenship.

Before the law was passed, members of Congress argued over the one-year requirement, wondered whether Jews and Catholics should be eligible for citizenship, and considered restrictions on the right of immigrants to hold political office. But no member publicly questioned the idea of limiting citizenship to only "free white persons."

Three years before the bill became law, Thomas Jefferson, the author of the Declaration of Independence, observed in his *Notes on the State of Virginia*, "It will probably be asked, Why not retain and incorporate the blacks into the state?" In response to that question, he advanced "as a suspicion only, that the blacks, whether originally a distinct race, or made distinct by time and circumstances are inferior to whites in the endowments both of body and mind." He called for "scientific investigations" but urged that researchers use caution "where our conclusion would degrade a whole race of men from the rank in the scale of beings which their Creator may perhaps have given them."[3]

Jefferson voiced his suspicions at a time when a growing number of Americans were urging that slavery be abolished. Their opposition was based in part on the ideas that Jefferson himself expressed in the Declaration of Independence. In response to his *Notes on the State of Virginia*, these abolitionists charged, "You have degraded the blacks from the rank which God hath given them in the scale of being! You have advanced the strongest argument for their state of slavery! You have insulted human nature!"

Some abolitionists offered Jefferson proof that people of African descent are equal to whites by citing the achievements of individuals like Benjamin Banneker, a free black from Maryland. The Georgetown (VA) *Weekly Ledger* described him in 1791 as "an Ethiopian whose abilities as surveyor and astronomer already prove that Mr. Jefferson's concluding that that race of men were void of mental endowment was without foundation."

Between 1791 and 1796, Banneker produced a series of almanacs—calendars containing weather forecasts, astronomical information, and other useful facts.

In the introduction to Banneker's first almanac, James McHenry, a prominent soldier and statesman, offered readers his personal assurance that Banneker had performed without assistance all of the mathematical calculations in the book. "I consider this Negro as fresh proof that the powers of the mind are disconnected with the color of the skin, or in other words, a striking contradiction to [the] doctrine that 'the Negroes are naturally inferior to the whites and unsusceptible of attainments in arts and sciences.'"

Shortly before publication, Banneker sent a hand-written copy of his almanac to Jefferson with a letter offering the book as evidence of what an individual of African descent could accomplish. In reply, Jefferson wrote, "Nobody wishes more than I do to see proofs as you exhibit, that nature has given to our black brethren, talents equal to those of the other colors of men, and that the appearance of a want of them is owing merely to the degraded condition of their existence, both in Africa and America."

Although Jefferson expressed admiration for Banneker's achievements, he continued to believe that blacks were inferior to whites. Nor did Banneker's almanac alter the way a growing number of other white Americans viewed people of African descent. By the early 1800s, even white Americans opposed to slavery increasingly regarded Africans as members of a separate and inferior race.

CONNECTIONS

How was *race* defined in Chapter 1? How is it defined in this reading? How do dictionaries define the term? What do you think it means to people in the United States today? What does it mean to you? How are these various definitions related to the word *equal*?

By the 1790s, slavery had existed in North America for nearly 200 years. How do you think the existence of slavery shaped the way Americans defined equality? Viewed race?

Jack Foley writes, "The only way for the 'majority' to conceive itself as a majority is to conceive of itself as white: without whiteness, there are only 'minorities.'" How does he seem to define "whiteness"? How do you define the term?

Every community has a "universe of obligation"—the name Helen Fein has given to the circle of individuals and groups "toward whom obligations are owed, to whom rules apply, and whose injuries call for [amends]."[4] Who was part of the nation's "universe of obligation" in the early years of the Republic? Who was excluded? What part did race play in definitions of citizenship?

How do you account for Jefferson's refusal to accept Banneker's accomplishments as proof of the abilities of African Americans? In your experience, what opinions are relatively easy to change? What opinions or impressions are more difficult to alter? What sorts of proof are most persuasive—personal experiences, the lessons of history, scholarly endorsements, philosophical arguments, scientific evidence—in changing an impression? Revising a stereotype? Altering a point of view?

Thomas Jefferson considered slavery immoral. Yet he was a slaveholder who saw Africans as a threat to "white racial purity." In reflecting on efforts to free the slaves, he wrote, "This unfortunate difference in color, and perhaps of faculty, is a powerful obstacle to the emancipation of these people." Despite such beliefs, Jefferson inspired generations of African Americans. In a speech, civil rights activist Julian Bond tried to explain why:

> Martin Luther King didn't care whether the . . . author of the Declaration of Independence thought he was inferior. The man may have thought so, but his words belied the thought. For King and his audiences, the significant Thomas Jefferson was not the Ambassador to France or the Secretary of State, the farmer or the slaveholder; as did Jefferson, they thought his chief virtue was as author of the Declaration of Independence, specifically of those self-evident truths that all are created equal. The promise of the words—for King, for those before him and us—became the true measure of the man.[5]

What is Bond suggesting about the power of ideas to spark the imagination and inspire creativity? Are Jefferson's most famous words the "true measure of the man" or should he be judged by his deeds? Why do you think some historians have called Jefferson's views paradoxical? To what extent did he seem to be aware of contradictions in his thinking? How did he try to resolve them?

---

1. The word *Negro* was commonly used in earlier centuries to refer to individuals of African descent. Its use reflects a particular time period.
2. "Multiculturism and the Media" by Jack Foley in *MultiAmerica: Essays on Cultural Wars and Cultural Peace*, edited by Ishmael Reed. Viking, 1997, pp. 367–369.
3. *Notes on the State of Virginia* by Thomas Jefferson, edited by William Peden. University of North Carolina Press, 1982, p. 143.
4. *Accounting for Genocide* by Helen Fein. Free Press, 1979, p. 4
5. "Address" by Julian Bond. Jefferson Conference, October 16, 1992, pp. 19-20.

Reading 3

Americans were debating the future of slavery and the role of African Americans in the nation at a time when scientists were trying to understand the world by naming, sorting, and categorizing every part of it. In the 1730s, Swedish naturalist Carolus Linnaeus devised a system that showed how living things are related to one another. Writer Jonathan Weiner notes that Linnaeus's system is often drawn as a "tree of life."

> The trunk of the tree divides near its base to form kingdoms, and each great trunk divides again and again into ever-finer branches and twigs; into species, subspecies, races, varieties, and, at last, like leaves on the twigs, individuals. We depict the order of life, in other words, as a family tree, a genealogy, in which the branches trace back to a common trunk. Every living thing is related, whether distantly or nearly, and every animal and plant shares the same ancestors at the root. . . .
>
> But that is not how Linnaeus himself saw his system. To him, and to other pious naturalists of his generation, . . . they represented the plan of God, who created the species in a single week, as described in the first pages of the Hebrew Bible: "And God created great whales . . . and every winged fowl after his kind: and God saw that it was good."
>
> . . . In Linnaeus's vast botanical collections he did notice many examples of local plant varieties, variations on a theme. But in his system these varieties were not half as significant as true species. . . . Local varieties were merely instances in which one of the Lord's created species had come to be adapted to its particular neighborhood.[1]

Linnaeus classified humankind as a species within the animal kingdom. He divided the human species into four varieties: European, American, Asiatic, and African. In his view, the four were more alike than different. By the late 1700s, a number of thinkers were trying to improve on Linnaeus's classification of humans.

In 1795, Johann Friedrich Blumenbach came up with a new classification scheme. In his book, *On the Natural Variety of Mankind*, he divided humanity into five varieties. As Linnaeus did, he associated each with a particular geographic area—Negro (African), Mongolian (Asian), Malay (Southeast Asia), American Indian (American), and Caucasian (European). Blumenbach introduced the word *Caucasian* "to describe the variety of mankind—the Georgian—that had originated on the southern slopes of Mount Caucasus." This, to Blumenbach, was the most beautiful race, and he said it must be "considered as the primate or intermediate of these five principal races." Other races represented "a degeneration from the original type."[2]

Although Blumenbach regarded Caucasians as the first and most beautiful variety of humans, he was careful to point out in *A Manual of the Elements of Natural History:*

> Although there seems to be so great a difference between widely separate nations, that you might easily take the inhabitants of the Cape of Good Hope, the Greenlanders, and the Circassians for so many different species of man, yet when the matter is thoroughly considered, you see that all do so run into one another, and that one variety of mankind does so sensibly pass into another that you cannot mark out the limits between them.

Like Blumenbach, Petrus Camper was also preoccupied with the idea of beauty and order in the world. Trained as an artist before turning to science, Camper was a professor of anatomy at the University of Groningen in the Netherlands. His interest in art and anatomy came together in the illustration on page 45, which originally appeared in a medical textbook printed in 1791, two years after his death.

Camper lived at a time when the Dutch were deeply involved in the international slave trade. Although Camper was personally opposed to slavery, he was fascinated by the stories and the artifacts brought home by sailors and merchants involved in the trade. He saw the skeletal remains of animals and humans from distant lands as pieces of a puzzle—each piece was a clue to a better understanding of the order of nature.

As a man of faith, Camper believed in monogenesis, the idea that all people share a common ancestry, even though, he thought that some groups had drifted further from the Biblical ideal than others. As a man of the Enlightenment, Camper believed that the world was ordered according to laws that could be discovered through reason and observation and then visually demonstrated. In such a world, he and others believed that an organism's "outer state"—its appearance—reflected its "inner state," its moral or intellectual worth.

Convinced that ancient Greece and Rome had come closer than other civilizations to perfection, he used Greek statues to establish standards of beauty. He ranked human faces by how closely they resembled this ideal. After measuring dozens of statues, Camper found that their "facial angle" averaged

100 degrees. (The facial angle is the angle formed by two intersecting lines—one drawn horizontally from the ears to the nose and other formed by the shape of the face from the upper lip to the forehead.) With this ideal in mind, Camper began measuring and sorting the skulls of apes and humans. He found that apes had a facial angle of 42 to 50 degrees. The average for the Europeans he measured was about 90 degrees and for Africans 70 degrees. (The intersecting lines on the drawing below indicate "facial angles.")

In the late 18th and early 19th centuries, a number of scientists ranked humankind along a "chain of being" based on Camper's facial angles. The idea of a "chain of being" dated back to the Middle Ages but gained new popularity in the years after Camper's death. As Kenan Malik explains in *The Meaning of Race*, "The Great Chain of Being linked the cosmos from the most miserable mollusk to the Supreme Being. Near the apex of this chain stood Man, himself graded by social rank. In this great chain, the humblest as well as the greatest played their part in preserving order and carrying out God's bidding."[3]

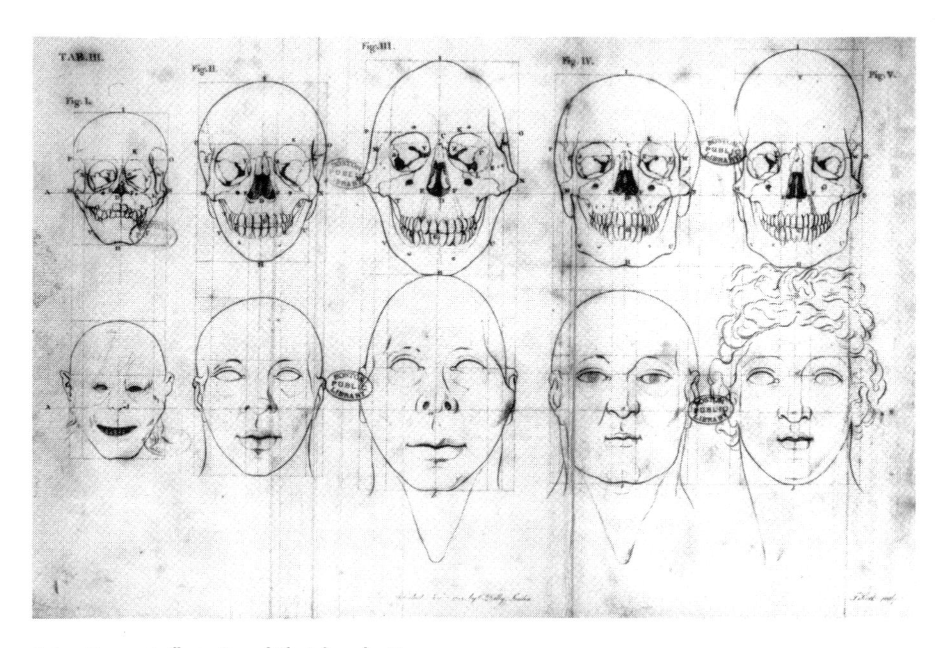

Petrus Camper's illustration of "facial angles."

CONNECTIONS

Linnaeus tested the idea that all living things are related to one another. What ideas were Blumenbach and Camper testing? To what extent were their methods good science—in the sense that they rigorously tested their hypotheses? To what extent did their approaches question "conventional wisdom"? To what extent did they reinforce conventional wisdom?

Linnaeus, Blumenbach, and Camper were all men of faith. How did their religious beliefs shape their observations of the natural world? What other aspects of their identity may have influenced the way they viewed differences among humankind? The value they placed on the similarities among humankind?

Why do you think Blumenbach regarded physical beauty as proof of superiority? Would he agree that "beauty is in the eye of the beholder"? How do notions of beauty affect the value we attach to individuals and groups?

Look carefully at Camper's illustration. If possible, project a slide of the image on a large screen and then discuss the illustration in small groups.

—Try not to explain the picture, simply describe what you notice. Have someone in the group record your observations and those of your classmates. You may also want to record your own impressions in your journal.

—Which faces look the most "human"? How does the artist use lines, shading, and shapes to convey a message? What characteristics make the drawings seem scientific? Authoritative?

—Based on your group's interpretation, give the drawing a title.

Camper called his drawing "The progression of skulls and facial expressions—from monkey, through black, to the average European and then thence to the Greek ideal-type." To what extent does his title support your impressions of the drawing? What is the significance of the word *progression*?

What kinds of proofs do you find more powerful—written proofs or visual evidence? Which is more likely to stretch the mind and inspire the imagination? Which is more difficult to forget? How do you think ideas like those of Blumenbach and Camper might have influenced people of the time? To what extent might the mystique of science keep the average person from questioning their ideas?

---

1. *The Beak of the Finch* by Jonathan Weiner. Random House, 1995, pp. 23–24.
2. Quoted in *Race and Manifest Destiny* by Reginald Horsman. Harvard University Press, 1981, p. 47.
3. *The Meaning of Race: Race, History and Culture in Western Society* by Kenan Malik. New York University Press, 1996, p. 43.

Reading 4

Petrus Camper believed in monogenesis, the idea that all people share a common ancestry based on the Biblical account of Adam and Eve. At the same time, he was convinced that some groups or "races" had declined further than others from their Biblical origin. He also suspected that there were intellectual and moral differences among the races as well as physical ones. In the mid-19th century, an American anthropologist, Samuel George Morton, extended Camper's work. But unlike Camper, Morton believed in polygenesis—the idea that each race was created separately. He also maintained that each race is fixed, intrinsically different from all others, and incapable of being changed.

Morton, a professor of medicine at the University of Pennsylvania, held two medical degrees and served as president of the Academy of Natural Sciences. According to the *New York Tribune*, "Probably no scientific man in America enjoyed a higher reputation among scholars throughout the world."[1] Like many scientists of his day, Morton believed that intelligence is linked to brain size. He therefore tried to rank the races according to skull size. After measuring a vast number of skulls from around the world, he concluded that whites have larger skulls than other races and are therefore "superior." He was not sure if blacks were a separate race or a separate species, but he did insist that people of African descent are different from and inferior to whites.

The following quotations are from Morton's *Crania Americana*, published in 1839. They suggest how physical differences can become markers that predict a group's intelligence, personality traits, even morality.

> **Europeans**
> The Caucasian Race is characterized by a naturally fair skin, susceptible of every tint; hair fine, long and curling, and of various colors. The skull is large and oval, and its anterior portion full and elevated. The face is small in proportion to the head, of an oval form, with well-proportioned features. . . . This race is distinguished for the facility with which it attains the highest intellectual endowments. . . .
>
> The spontaneous fertility of [the Caucasus] has rendered it the hive of many nations, which extending their migrations in every direction, have peopled the finest portions of the earth, and given birth to its fairest inhabitants. . . .
>
> **Asians**
> This great division of the human species is characterized by a sallow

or olive colored skin, which appears to be drawn tight over the bones of the face; long black straight hair, and thin beard. The nose is broad, and short; the eyes are small, black, and obliquely placed, and the eye-brows are arched and linear; the lips are turned, the cheek bones broad and flat. . . . In their intellectual character the Mongolians are ingenious, imitative, and highly susceptible of cultivation [i.e. learning].

So versatile are their feelings and actions, that they have been compared to the monkey race, whose attention is perpetually changing from one object to another. . . .

Native Americans

The American Race is marked by a brown complexion; long, black, lank hair; and deficient beard. The eyes are black and deep set, the brow low, the cheek-bones high, the nose large and aquiline, the mouth large, and the lips tumid [swollen] and compressed. . . . In their mental character the Americans are averse to cultivation, and slow in acquiring knowledge; restless, revengeful, and fond of war, and wholly destitute of maritime adventure.

They are crafty, sensual, ungrateful, obstinate and unfeeling, and much of their affection for their children may be traced to purely selfish motives. They devour the most disgusting [foods] uncooked and uncleaned, and seem to have no idea beyond providing for the present moment. . . . Their mental faculties, from infancy to old age, present a continued childhood. . . . [Indians] are not only averse to the restraints of education, but for the most part are incapable of a continued process of reasoning on abstract subjects. . . .

Africans

Characterized by a black complexion, and black, woolly hair; the eyes are large and prominent, the nose broad and flat, the lips thick, and the mouth wide; the head is long and narrow, the forehead low, the cheek-bones prominent, the jaws protruding, and the chin small. In disposition the Negro is joyous, flexible, and indolent; while the many nations which compose this race present a singular diversity of intellectual character, of which the far extreme is the lowest grade of humanity. . . .

The moral and intellectual character of the Africans is widely different in different nations. . . . The Negroes are proverbially fond of their amusements, in which they engage with great exuberance of spirit; and a day of toil is with them no bar to a night of revelry.

> Like most other barbarous nations their institutions are not infrequently characterized by superstition and cruelty. They appear to be fond of warlike enterprises, and are not deficient in personal courage; but, once overcome, they yield to their destiny, and accommodate themselves with amazing facility to every change of circumstance.
>
> The Negroes have little invention, but strong powers of imitation, so that they readily acquire mechanic arts. They have a great talent for music, and all their external senses are remarkably acute.[2]

Morton's ranking of the "races" had very real consequences. After meeting Morton and viewing his skull collection, Louis Agassiz, a noted biologist who joined the faculty of Harvard University in 1846, taught his students that Africans are a separate species. In evaluating Agassiz's career, anthropologist Lee D. Baker observes: "Agassiz's legacy is not only the statues, schools, streets, and museums in Cambridge [Massachusetts] emblazoned with his name but also the bevy of students who were under his tutelage at Harvard University. He trained virtually all of the prominent U.S. professors of natural history during the second half of the nineteenth century."[3]

Morton's rankings also shaped the way many politicians, journalists, and ministers viewed two of the most pressing social and political issues of the day: the expulsion of Native Americans from their ancestral lands and the expansion of slavery. Between 1816 and 1850, over 100,000 Indians from 28 tribes were forced from their homes east of the Mississippi to western lands that white Americans considered useless. At the same time, about 3.5 million African Americans were held in bondage. Their enslavement prompted a heated debate between slave-owners and an international community of abolitionists, opponents to slavery. Morton's writings played a part in both debates by promoting the idea that the Constitution does not apply to Native Americans or Africans because they are not the sorts of people for whom the document was written.

CONNECTIONS

Camper believed in monogenesis and Morton in polygenesis. How did those beliefs shape the way each viewed differences?

List the adjectives Morton uses to define each of the four groups. Circle every adjective that has a positive connotation. Is there a correlation between the number of positive adjectives that Morton uses in describing a group and his estimate of its moral or intellectual "worth"?

What do you think Morton meant when he wrote that Africans "yield to their destiny and accommodate themselves" to new circumstances? What does the word *destiny* imply? How might Morton's writings influence debates over slavery? How might they justify the removal of Native Americans to remote areas?

What power do teachers have to shape the way their students view the world? What power do parents have? A community? Religious leaders? To what extent did teachers like Morton and Agassiz betray their students?

The link Morton and others saw between brain size and intelligence shaped ideas about not only African Americans and Native Americans but also women. In 1879, Gustave Le Bon, a French student of anthropology, wrote:

> In the most intelligent races, as among the Parisians, there are a large number of women whose brains are closer in size to those of gorillas than to the most developed male brains. This inferiority is so obvious that no one can contest it for a moment; only its degree is worth discussion. . . . Without doubt there exist some distinguished women, very superior to the average man, but they are as exceptional as the birth of any monstrosity, as, for example, of a gorilla with two heads, consequently, we may neglect them entirely. . . . A desire to give them the same education, and as a consequence to propose the same goals for them is a dangerous [illusion].[4]

Define the word *scientific*. Are scientific proofs more convincing than other proofs? How difficult are they to counter? For example, how might a woman "prove" that she is the equal of a man? How do you think Le Bon would respond to her proof? How might an African American "prove" that he or she is the equal of any other American? How do you think Morton, Le Bon, and others would respond to that proof?

---

1. Quoted in *The Leopard's Spots: Scientific Attitudes Towards Race in America, 1815–1859* by W. Stanton. University of Chicago Press, 1960, p. 144.
2. *Crania Americana* by Samuel George Morton. John Pennington, 1839, pp. 5, 6, 50, 54, 81.
3. *From Savage to Negro: Anthropology and the Construction of Race, 1896-1954* by Lee D. Baker. University of California Press, 1998, p.16.
4. Quoted in *The Mismeasure of Man* by Stephen Jay Gould. W.W. Norton, 1981, pp. 104–105.

Reading 5

Samuel Morton saw himself as an impartial scientist with no interest in partisan politics. He insisted that his conclusions were based solely on the results of his scientific investigations. In 1849, Morton summarized his work in a paper entitled "Observations on the Size of the Brain in Various Races and Families of Man." The table below is taken from that paper.

TABLE,

Showing the Size of the Brain in cubic inches, as obtained from the internal measurement of 623 Crania of various Races and Families of Man.

RACES AND FAMILIES.	No. of Skulls.	Largest I. C.	Smallest I. C.	Mean.	Mean.
MODERN CAUCASIAN GROUP.					
TEUTONIC FAMILY.					
Germans,	18	114	70	90	92
English,	5	105	91	96	
Anglo-Americans.	7	97	82	90	
PELASGIC FAMILY. *Persians, Armenians, Circassians.*	10	94	75	84	
CELTIC FAMILY. *Native Irish.*	6	97	78	87	
INDOSTANIC FAMILY. *Bengalees, &c.*	32	91	67	80	
SEMITIC FAMILY. *Arabs.*	3	98	84	89	
NILOTIC FAMILY. *Fellahs.*	17	96	66	80	
ANCIENT CAUCASIAN GROUP.					
From the Catacombs. PELASGIC FAMILY. *Græco-Egyptians.*	18	97	74	88	
From the Catacombs. NILOTIC FAMILY. *Egyptians.*	55	96	68	80	
MONGOLIAN GROUP.					
CHINESE FAMILY.	6	91	70	82	
MALAY GROUP.					
MALAYAN FAMILY.	20	97	68	86	85
POLYNESIAN FAMILY.	3	84	82	83	
AMERICAN GROUP.					
TOLTECAN FAMILY.					79
Peruvians,	155	101	58	75	
Mexicans.	22	92	67	79	
BARBAROUS TRIBES. *Iroquois, Lenapé, Cherokee, Shoshoné, &c.*	161	104	70	84	
NEGRO GROUP.					
NATIVE AFRICAN FAMILY.	62	99	65	83	83
AMERICAN-BORN NEGROES.	12	89	73	82	
HOTTENTOT FAMILY.	3	83	68	75	
ALFORIAN FAMILY. *Australians.*	8	83	63	75	

Notice that the number of skulls varies from group to group. Morton measured "cranial capacity"—the interior size of the skull—in cubic inches.

After reading Samuel Morton's *Crania Americana*, Frederick Douglass, a leader in the fight against slavery and himself a former slave, strongly disagreed. He described the scientist as reasoning "from prejudice rather than from facts." Douglass went on to say: "It is the province of prejudice to blind; and scientific writers, not less than others, write to please, as well as to instruct, and even unconsciously to themselves, (sometimes,) sacrifice what is true to what is popular. Fashion is not confined to dress; but extends to philosophy as well—and it is fashionable now, in our land, to exaggerate the differences between the Negro and the European."[1]

In the mid-1800s, however, most scientists accepted both the methods and the data Morton used to arrive at his conclusions. Among the few to raise questions was Friedrich Tiedemann, a German professor who also used skulls to investigate the ways race, intelligence, and brain size are linked. According to *The Skull Measurer's Mistake* by Sven Lindqvist, Tiedemann measured skulls by filling them with millet, then weighing the millet. The largest skull in his collection held 59 ounces. It was from a Native American man. The second largest was from a white man. Third was an African, fourth a white, and fifth place was shared by three whites, a Mongol, and a Malay. The largest female skull came from a Malay woman, whose cranium held 41 ounces. A white and a Native American shared second place, and one black and one white woman shared third place. These results did not support Morton's conclusion. Nor did they fit the race hierarchy of Tiedemann's time, in which whites were always at the top and blacks at the bottom.

CONNECTIONS

What does it mean for a human being to be measured, ranked, and then labeled? Does it matter who does the measuring? What is the difference between being ranked by a scientist or a teacher? By a relative or a friend? Which has a greater impact on the way you see yourself and others?

What is the scientific method? Why do scientists claim it is impartial? How important is impartiality to the method?

To evaluate a scientific finding, it is important that the data be not only accurate but also relevant to the question under investigation. How would you determine the accuracy of Morton's data? How relevant are Morton's measurements to the qualities he ascribes to the various "racial groups" (Reading 4)?

The *mean* is the average—the sum of all of the items divided by the number of items or in this case, the number of skulls. What conclusions about cranial capacity does the "Mean" column in Morton's table suggest?

In *The Mismeasure of Man*, biologist Stephen Jay Gould uses Morton's original notes and raw data to evaluate his methods. Gould concludes:

—Morton's sub-samples were not inclusive. For example, of the 333 skulls in his "American Indians" sample, 155 were Inca from South America. Their skulls tended to be smaller than those of other Indian groups. At the same time, he lumped the relatively larger Iroquois skulls into a separate category called "Barbarous Tribes." When he found that skulls from India were smaller than other Caucasian skulls, he omitted them from his "Modern Caucasian" group.

—Morton's measurements were influenced by his subjective expectations. Morton used mustard seed to measure the cranial capacities of his skulls. Gould found that the seeds were often packed tightly in the European skulls but not in Indian or African American skulls. As a result, Morton inflated the sizes of European skulls and deflated those of other groups.

—Morton failed to correct his figures for gender and stature. Since females tend to be smaller than males, they have smaller skulls. Morton included more female skulls in his African and Indian groups than in his European group. The result was to inflate the size of European skulls and decrease those of other groups.

—Morton miscalculated some numbers and left out others. For example, he rounded down measurements for Egyptian skulls and rounded up measurements of German and Anglo-Saxon skulls.[2]

Despite these errors, Gould does not think that Morton intended to deceive anyone. If that had been his intention, he probably would have tried to cover up his data and hide his procedures. How would you account for Morton's errors? What does Gould's study suggest about the ways unconscious assumptions may affect one's objectivity?

To illustrate some of the problems Gould sees in Morton's measurements, pour the contents of a bag of peppercorns into a skull (plastic or real) until it seems full. Then pour the peppercorns into a calibrated beaker. Record the volume and round the number downward. Then redo the experiment. This time try to get as many peppercorns into the skull as possible. Pour the new amount into the beaker and round the number up instead of down. Same skull—two different calculations!

What similarities do you notice in the ways Morton and Tiedemann approached their research? What differences seem most striking? How important are those differences?

In the 1870s, Paul Broca, a noted French anthropologist, criticized Tiedemann's research as "imprecise." He believed that Tiedemann had "set out to prove that the cranial capacity of all human races is the same." Yet Broca had nothing but praise for Morton. What does the word *objective* mean? To what extent was Morton objective? Tiedemann? Identify instances where unstated assumptions have affected your objectivity or the objectivity of someone you know.

Tiedemann came to believe that there is no relationship between skull size and intelligence as a result of his study of dolphins. He found that the size of a dolphin's brain varies with gender, body length, body weight, and body condition but not with intelligence. His work with dolphins also taught Tiedemann "not only to study averages in large groups, but also to take an interest in individuals and the variations between individuals as well as between groups and divisions of groups." What may a study of averages in large groups reveal? What may it conceal? What may a study of individuals and the variations among them reveal? What may such a study conceal?

Both Tiedemann and Douglass challenged the "conventional wisdom" about race. What experiences may have prompted them to question ideas? Challenge assumptions? What is the conventional wisdom about "race" today? Who determines the conventional wisdom? How does the "conventional wisdom" change?

---

1. *The Life and Writings of Frederick Douglass*, Vol. 2. Edited by Philip S. Foner. International Publishing, 1950, p. 298.
2. Paraphrased from *Mismeasure of Man* by Stephen Jay Gould. W.W. Norton & Co., 1996, 1981, pp. 100-101.

Reading 6

By the middle of the 1800s, the idea that some "races" are superior to others had become the "conventional wisdom." Respected scientists like Samuel Morton gave racism legitimacy. As a result, racist ideas were taught in universities, preached from pulpits, and reinforced in books, magazines, and newspapers. After surveying the leading publications of the day, historian Reginald Horsman notes, "One did not have to read obscure books to know that the Caucasians were innately superior, and that they were responsible for civilization in the world, or to know that inferior races were destined to be overwhelmed or even to disappear. These ideas permeated the main American periodicals and in the second half of the century formed part of the accepted truth of America's schoolbooks." [1] They also shaped the way Americans defined citizenship.

Immediately after the American Revolution, only three states—Virginia, South Carolina, and Georgia—limited the right to vote to white men. Until 1800, no northern state limited suffrage on the basis of race. After 1800, however, every state that entered the Union with the exception of Maine placed restrictions on the right of African Americans to vote. States that permitted blacks to vote began to narrow or remove that right entirely. In 1837, a delegate to the Pennsylvania Constitutional Convention justified taking away voting rights from African American citizens by describing the United States as "a political community of white persons." By the late 1850s, blacks could vote on the same basis as whites only in five states—all of them in New England.

In 1857, the language of exclusion reached the Supreme Court. In the Dred Scott decision, Chief Justice Roger B. Taney ruled that blacks "had no rights which the white man was bound to respect." The American people, Taney argued, constituted a "political family" restricted to whites. Historian Eric Foner notes, "It was a family of which blacks, descended from different ancestors and lacking a history of freedom, could never be a part. In effect, race had replaced class as the boundary separating which American men were entitled to enjoy political freedom and which were not."[2]

As race increasingly defined citizenship, free blacks in the North and West as well as the South found themselves outside the nation's "universe of obligation." When they looked for work on the docks of New York City, they were attacked by white workers. When young African Americans applied for the apprenticeships that would lead to good jobs in places like Cincinnati, Ohio, white mechanics blocked their every attempt. As one young black man complained, "Why should I strive hard and acquire all the constituents of a man, if the prevailing genius of the land admit me not as such, or but in an inferior degree!

Pardon me if I feel insignificant and weak. . . . What are my prospects? To what shall I turn my hand? Shall I be a mechanic? No one will employ me; white boys won't work with me. . . . Drudgery and servitude, then, are my prospective portion."[3]

Even the most educated African Americans experienced hostility, prejudice, and discrimination at every turn. Historian Ronald Takaki relates the experiences of Martin Delany, the son of a slave father and free mother in Charles Town, Virginia (now Charleston, West Virginia), to suggest the breadth and depth of the shame and humiliation African Americans experienced in all parts of the nation in the early 1800s.

> As a child, Martin learned that his membership in the black race made him the object of white scorn. [His mother's] efforts to teach her children to read and write aroused angry opposition from white neighbors who were anxious to preserve their belief in black intellectual inferiority. . . . White resentment was so intense that she felt compelled to move her family across the border to Pennsylvania.
>
> But even north of slavery, racism was prevalent. As a young man studying in Pittsburgh during the 1830s, Delany experienced the brutality of anti-black riots led by mobs composed of white workers.
>
> As a journalist and as an antislavery lecturer during the 1840s, Delany traveled widely throughout the North and often encountered racial hostility and violence. On one occasion, a white mob in Marseilles, Ohio, threatened to tar and feather him and burn him alive. Delany found that white children, even while involved in play, were never too busy to notice a black passing by and scream "nigger.". . . Delany found that the racial epithets were not only "an abuse of the feelings," but also "a blasting outrage on humanity."
>
> His bitterness toward northern society was sharpened by an admissions controversy at Harvard Medical School. In 1850, Delany along with two other blacks were admitted to the school. Their admission, however, was conditional: upon graduation, they would have to emigrate and practice medicine in Africa. Even so, their presence at Harvard provoked protests from white students. Demanding the dismissal of the blacks, they argued that integration would lower the "reputation" of Harvard and "lessen the value" of their diploma. The whites refused to attend classes with the blacks. . . .
>
> The faculty quickly capitulated, ignoring a student counter-petition favoring the admission of the blacks. Deeming it "inexpedient" to allow blacks to attend lectures, the faculty defended their decision based on their commitment to teaching and academic excellence.[4]

Two years after the incident at Harvard, Delany wrote a book that encouraged African Americans to return to Africa. He was convinced that even if slavery were abolished, blacks would not be accepted as equals in the United States. Yet even as he made plans to leave the country, he dedicated his book "to the American people, North and South. By their most devout and patriotic fellow-citizen, the author." He also reminded his readers of the contributions that blacks had made to the nation. "Among the highest claims that an individual has upon his country," he wrote, "is that of serving in its cause, and assisting to fight its battles." In 1861, when the Civil War began, he abandoned his dreams of Africa and volunteered for the Union Army. He served as a major in the 104th Regiment of the United States Colored Troops.

CONNECTIONS

According to scholar Leon Higginbotham, Jr., race was increasingly entwined with the idea of citizenship in the years just after the American Revolution. Increasingly, he writes, a citizen was a man who could help his neighbors put down slave rebellions or fight Indians. How is the word *citizen* defined by the mid-1800s? How did notions about race shape that definition?

A young African American quoted in this reading asks, "What are my prospects? To what shall I turn my hand?" How do you think Samuel Morton and other "race scientists" would answer his questions? How might Martin Delany answer them? How would you answer them? To what extent do Samuel Morton's rankings place that young man and other African Americans beyond the nation's "universe of obligation"? What does Martin Delany's story suggest about the consequences of being outside a nation's universe of obligation?

How do you explain the change from a society that emphasizes equality to one that stresses differences? What role may education have played in that change?

---

1. *Race and Manifest Destiny* by Reginald Horsman. Harvard University Press, 1981, p. 157.
2. *The Story of American Freedom* by Eric Foner. W.W. Norton & Company, 1998, pp. 74–75.
3. Quoted in *North from Slavery* by Leon Litvak. Chicago, 1965, pp. 153-154.
4. *A Different Mirror* by Ronald Takaki. Copyright © 1993 by Ronald Takaki. By permission of Little, Brown and Company, 1993, pp. 127-128.

Reading 7

When Thomas Jefferson questioned the intellectual capabilities of people of African descent in the late 1700s, his opponents reminded him of the words he wrote in the Declaration of Independence: "We hold these truths to be self-evident: That all men are created equal; that they are endowed by their Creator with certain inalienable rights; that among these are life, liberty, and the pursuit of happiness." By the mid-1800s, Governor James Hammond of South Carolina and a growing number of other white Americans viewed "as ridiculously absurd, that much lauded but nowhere accredited dogma of Mr. Jefferson that 'all men are born equal.'"

Among the few Americans in the early 1800s to keep alive the language of the Declaration of Independence were abolitionists—those who sought to end slavery in the nation. Although many of them did not believe that "all men are born equal," their long struggle to abolish slavery gave new meaning to personal liberty and the rights attached to citizenship.

In the 1830s, writes historian Eric Foner, politicians and ordinary citizens tried to silence those who were critical of slavery. In northern cities, mobs broke up the meetings of abolitionist societies and destroyed their printing presses. In 1836, the U.S. House of Representatives refused to consider any petition that called for the abolition of slavery. At about the same time, Postmaster General Amos Kendall allowed U.S. postal officials in southern states to remove from the mail any written material critical of slavery. Foner argues:

> The fight for the right to debate slavery openly and without reprisal led abolitionists to elevate "free opinion"—freedom of speech and of the press and the right of petition—to a central place in what [William Lloyd] Garrison called the "gospel of freedom." The struggle for free speech also reinforced the contention that slavery threatened the liberties of white Americans as well as black. Free expression, abolitionists insisted, should be a national standard, not subject to limitation by those who held power within local communities.[1]

The struggle against slavery also inspired two definitions of citizenship. One was based on race. The other was based on a civic understanding of nationhood. It was summarized by Lydia Maria Child in 1833 in a popular essay entitled "An Appeal in Favor of that Class of Americans Called Africans." Foner writes:

> Child's text insisted that blacks were compatriots, not foreigners; they were no more Africans than whites were Englishmen. At a time

> when the authority to define the rights of citizens lay almost entirely with the states, abolitionists maintained that "birth place" should determine who was an American. The idea of birthright citizenship, later enshrined in the Fourteenth Amendment, was a truly radical departure from the traditions of American life.[2]

Black abolitionists were particularly adamant in their insistence on the equality of African Americans. In 1854, in a speech in Cleveland, Ohio, Frederick Douglass responded to an editorial in a Virginia newspaper that justified slavery by claiming that African Americans were less human than white Anglo Saxons—the descendents of a mythical people who settled in England in the fifth century. Douglass told his audience:

> Man is distinguished from all other animals by the possession of certain definite faculties and powers, as well as by physical organization and proportions. He is the only two-handed animal on the earth—the only one that laughs, and nearly the only one that weeps. Men intuitively distinguish between men and brutes. Common sense itself is scarcely needed to detect the absence of manhood in a monkey, or to recognize its presence in a Negro. His speech, his reason, his power to acquire and to retain knowledge, his heaven-erected face, his [inclinations], his hopes, his fears, his aspirations, his prophecies plant between him and the brute creation a distinction as eternal as it is palpable. Away, therefore, with all the scientific moonshine that would connect men and monkeys; that would have the world believe that humanity, instead of resting on its own characteristic pedestal—gloriously independent—is a sort of sliding scale, making one extreme brother to the orangutan, and the other to angels, and all the rest intermediates!
>
> Tried by all the usual, and all the unusual tests, whether mental, moral, physical, or psychological, the Negro is a MAN—considering him as possessing knowledge, or needing knowledge, his elevation or his degradation, his virtues, or his vices—whichever road you take, you reach the same conclusion, the Negro is a MAN. His good and his bad, his innocence and his guilt, his joys and his sorrows, proclaim his manhood in speech that all mankind practically and readily understand.
>
> A very [profound] author says that "man is distinguished from all other animals, in that he resists as well as adapts himself to his circumstances." He does not take things as he finds them, but goes to work to improve them. Tried by this test, too, the Negro is a man. You may see him yoke the oxen, harness the horse and hold the plow. He

can swim the river; but he prefers to fling over it a bridge. The horse bears him on his back—admits his mastery and dominion. The barnyard fowl know his step, and flock around to receive their morning meal from his sable hand. The dog dances when he comes home, and whines piteously when he is absent. All these know that the Negro is a MAN. Now, presuming that what is evident to beast and to bird, cannot need elaborate argument to be made plain to men, I assume, with this brief statement, that the Negro is a man.

. . . Indeed, ninety-nine out of every hundred of the advocates of a diverse origin of the human family [i.e., polygenesis, or multiple creations] in this country, are among those who hold it to be a privilege of the Anglo-Saxon to enslave and oppress the African—and slaveholders, not a few, like the Richmond Examiner to which I have referred, have admitted, that the whole argument in defense of slavery, becomes utterly worthless the moment the African is proved to be equally a man with the Anglo-Saxon. The temptation therefore, to read the Negro out of the human family is exceeding strong, and may account somewhat for the repeated attempts on the part of Southern pretenders to science, to cast a doubt over the Scriptural account of the origin of mankind. . . .

By making the enslaved a character fit only for slavery, [slaveholders] excuse themselves for refusing to make the slave a freeman. A wholesale method of accomplishing this result, is to overthrow the instinctive consciousness of the common brotherhood of man. For, let it be once granted that the human race are of multitudinous origin, naturally different in their moral, physical, and intellectual capacities, and at once you make plausible a demand for classes, grades, and conditions, for different methods of culture, different moral, political, and religious institutions, and a chance is left for slavery, as a necessary institution.[3]

CONNECTIONS

Nations, like individuals, have an identity. Make an identity chart for the United States in 1776. What values and beliefs were central to the nation's identity? What changes were Americans making in that chart in the early 1800s? In 1860? What might such a chart look like today?

What motive does Douglass attribute to those who want to "read the Negro out of the human family"? What does he consider the logical result of a belief in polygenesis?

Sociologist Orlando Patterson writes that the "first men and women to struggle for freedom, the first to think of themselves as free in the only meaningful sense of the term were freedmen." What is the paradox, or seeming contradiction, Patterson describes?

In 1861, the United States fought a civil war over the right of African Americans to be free—to be part of the "human family." When the Civil War ended in 1865, the nation added three amendments to the Constitution. The Thirteenth Amendment ended slavery. Research the Fourteenth and Fifteenth amendments. What was the goal of each amendment? Why do many constitutional experts regard the Fourteenth as the more revolutionary of the two?

People often think of a historical event in terms of a simple cause and an immediate effect. How does the long crusade against slavery complicate that view? To more fully appreciate its legacies, you may want to investigate the history of the Civil Rights Movement or of particular groups like the National Association for the Advancement of Colored People or the American Civil Liberties Union.

The editorial to which Douglass responded appeared in a Virginia newspaper. If Douglass had expressed his ideas in a letter to the editor, it would not have been published. If he had given his speech in the state of Virginia rather than Ohio, he would have been arrested. By the mid-1800s, Virginia and almost every other southern state outlawed anti-slavery publications and speeches. What effect do you think such limitations on debate had on science and scientists? On democracy? What is the link between scientific inquiry and freedom of expression? Between democracy and freedom of expression?

---

1. *The Story of American Freedom* by Eric Foner. W.W. Norton & Company, 1998, pp. 85-86.
2. Ibid., p. 86
3. "The Claims of the Negro Ethnologically Considered" by Frederick Douglass. Excerpts from an address delivered at Western Reserve College, July 12, 1854.

3. Evolution, "Progress," and Eugenics

Scientific writers, not less than others, write to please, as well as to instruct, and even unconsciously to themselves, (sometimes,) sacrifice what is true to what is popular.

Frederick Douglass

Chapter 2 explored the effects of "race science" on the way Americans viewed differences in the years before the Civil War. This chapter focuses on the impact of a new scientific theory published in England just before the war began—Charles Darwin's theory of evolution. Noting that living things change from generation to generation, Darwin argued that new forms of life eventually develop from (and sometimes replace) old forms. In the decades after the Civil War, scholars applied that theory to not only the natural world but also human societies. It seemed to "explain" all of the differences they observed in the world.

By 1900, writes historian Page Smith, Darwin's theory "colors the way social classes view themselves and, more important, the way they view other classes. It affects attitudes toward other races . . . especially American Indians, blacks, [East Asians], all of whom are generally viewed as representing lower stages of evolutionary development. It is taken by some Americans, generally wealthy and 'successful,' as confirming the model of competitive individualism and thereby justifying capitalism, and it is taken by many others as anticipating socialism as a higher and more humane form of political and economic organization. It divides clerics and professors of philosophy, natural scientists and 'social scientists,' husbands and wives, parents and children."[1]

In time, some thinkers came to believe that evolution could do more than explain physical and social differences. It could be used to "improve the race" through eugenics—a new branch of scientific inquiry developed by Darwin's cousin, Francis Galton. He claimed that eugenics would "raise the present miserably low standard of the human race" by "breeding the best with the best." At another time, that idea might have been dismissed or ignored. In the early 1900s, many people found it appealing. What attracted them to eugenics? Was it "good science" or, as Frederick Douglass once argued, another example of scientists "sacrificing what is true to what is popular"? Chapter 3 addresses these questions. Many of the readings suggest what can happen when unexamined ideas about difference are used to justify social inequalities, deny opportunities, and legitimize discrimination. They also explore the complicated relationship between science and society.

---

1. *The Rise of Industrial America* by Page Smith. McGraw-Hill, 1984, p. xiii.

From Darwin to Social Darwinism

Reading 1

Charles Darwin's theory of evolution grew out of a journey he made to South America on a survey ship. Between 1831 and 1836, while the crew mapped the coast of South America, the young Englishman collected plants and animals at every stop. The sights he saw on the voyage and the specimens he gathered transformed the way he viewed the natural world. In time, his vision would also alter the way people everywhere saw themselves and others.

His observations convinced Darwin that species develop in different directions when they are isolated from one another. But he did not have any idea of how it happened until he sat down one evening to read *An Essay on the Principle of Population as It Affects the Future Improvement of Society* by the Reverend Thomas Malthus. According to Malthus, human populations multiply faster than the supply of food. If that is also true of animals, Darwin reasoned, they must compete to stay alive. So it is nature that "selects" the forms of life most likely to survive. "Here then I had at last got a theory by which to work!" he wrote.

Darwin concluded that all living things struggle to obtain food, water, and a safe habitat. An organism that is well suited to its environment has the best chance of living long enough to mate and produce offspring. Gradually, as some organisms thrive and others die out, new traits, species, and forms of life develop or evolve. Darwin called this process "natural selection." In 1859, Darwin published his theories in a book entitled *On the Origin of Species*. It became an almost instant sensation.

Many readers immediately saw connections between Darwin's theory of evolution and their own society. A number of them were influenced by the writings of Herbert Spencer, a British thinker. Referring to Darwin's work but using his own phrases such as "the struggle for existence" and "the survival of the fittest," Spencer helped popularize a doctrine known as "social Darwinism."

In every country, people interpreted social Darwinism a little differently. In Germany, Ernst Haeckel, a biologist, combined the doctrine with romantic ideas about the German people. In a book called *The Riddle of the Universe*, he divided humankind into races and ranked each. "Aryans" were at the top of his list and Jews and Africans at the bottom. In the United States and England, social Darwinists stressed the idea that competition rewards "the strong." As a result, many of them opposed aid to the poor, laws that would place limits on cut-throat competition, and efforts to regulate working conditions in the nation's factories. They wanted government to let nature take its course.

Spencer and his followers argued that individuals and groups who undertake "in a wholesale way to foster good-for-nothings" commit an "unquestionable injury" by stopping "that natural process of elimination by which society continually purifies itself."[1] William Graham Sumner, a professor at Yale and a follower of Spencer, explained further:

> Every man and woman in society has one big duty. That is, to take care of his or her own self. This is a social duty. For, fortunately, the matter stands so that the duty of making the best of one's self, individually, is not a separate thing from the duty of filling one's place in society, but the two are one. . . .
>
> Now the man who can do anything for or about anybody else than himself is fit to be head of a family; and when he becomes head of a family he has duties to his wife and his children, in addition to the former big duty. . . . If, now, he is able to fulfill all this, and to take care of . . . his family and his dependents, he must have a surplus of energy, wisdom, and moral virtue beyond what he needs for his own business. No man has this; for a family is a charge that is capable of infinite development, and no man could suffice to the full measure of duty for which a family may draw upon him. . . .
>
> Society, therefore, does not need any care or supervision. If we can acquire a science of society, based on observation of phenomena and study of forces, we may hope to gain some ground slowly toward . . . a sound and natural social order.[2]

Not surprisingly, social Darwinism had special appeal for the rich and powerful. To them, it seemed to explain inequalities among not only individuals but also social classes and races. Some social Darwinists combined Samuel Morton's racial hierarchy (pages 47-49) with the theory of natural selection to create a new "more scientific" way of justifying prejudice and discrimination. These theories appealed to many white Americans, including a number of religious leaders. The Reverend Josiah Strong was one of the most influential writers in the late 1800s. In 1885, he wrote:

> There is apparently much truth in the belief that the wonderful progress of the United States, as well as the character of the people, are the results of natural selection; for the more energetic, restless and courageous men from all parts of Europe have emigrated during the last ten or twelve generations to that great country, and have there succeeded best. Looking to the distant future, I do not think that the Reverend Mr. Zincke takes an exaggerated view when he says: "All other series of events—as that which resulted in the culture of mind in

Greece and that which resulted in the empire of Rome—only appear to have purpose and value when viewed in connection with, or rather subsidiary to . . . the great stream of Anglo-Saxon emigration to the west." [3]

Once the West was settled, Strong declared:

Then will the world enter upon a new stage of its history—the final competition of races for which the Anglo-Saxon is being schooled. If I do not read amiss, this powerful race will move down upon Mexico, down upon Central and South America, out upon the islands of the sea, over upon Africa and beyond. And can anyone doubt that the result of this competition of races will be the "survival of the fittest"? [4]

In 1896, the United States Supreme Court expressed a view similar to Strong's in deciding a case known as *Plessy v. Ferguson*. Homer Plessy, an African American, challenged a Louisiana law that kept blacks separated from whites on public transportation. He argued that John Ferguson, the Louisiana judge who convicted him, had violated his rights as stated in the Fourteenth Amendment to the United States Constitution. That amendment guarantees every citizen equal protection under the law. Eight of the nine justices sided with Ferguson, who argued that as long as the railroad offered "separate but equal" seating for whites and blacks, Plessy's rights were protected. In expressing the majority opinion, Associate Justice Henry B. Brown asserted, "If one race be inferior to the other socially, the Constitution of the United States cannot put them on the same plane."

The decision permitted the growth of a system of state and local legislation known as "Jim Crow" laws. They established racial barriers in almost every aspect of American life. In many places, black and white Americans could not publicly sit, drink, or eat side by side. Churches, theaters, parks, even cemeteries were segregated. By the early 1900s, writes historian Lerone Bennett, Jr., "America was two nations—one white, one black, separate and unequal." He likens segregation to "a wall, a system, a way of separating people from people." That wall, which did not go up in a single day, was built "brick by brick, bill by bill, fear by fear."[5]

That wall shaped the opportunities open to African American children throughout the nation, but most particularly in the South. Historian Leon F. Litwack writes:

When Pauli Murray entered school in Durham, North Carolina, in the 1920s, she inherited nearly half a century of separate and unequal education in the South. The schoolhouse, located in the West End, resembled a warehouse more than a school. The dilapidated two-story wooden structure creaked and swayed in the wind as if it might collapse. The exterior showed the effects of some hard winters. The interior featured bare and splintery floors, leaky plumbing, broken drinking fountains, and smelly toilets that were usually out of order. "It was never the hardship which hurt so much," Murray remembered, "as the contrast between what we had and what the white children had." The new books white children used ("we got the greasy, torn, dog-eared books"), the field days in the city park the white children enjoyed ("we had it on a furrowed stubby hillside"), the prominent mention white children received in the newspapers ("we got a paragraph at the bottom"), the attention bestowed on public displays in white schools by city officials, including the mayor ("we got a solitary official")—all served to set the white schools apart from the black schools. No one pretended to take seriously the Supreme Court decision commanding separate but equal schools. To Murray, the school she attended defined her very being. "Our seedy, rundown school told us that if we had any place at all in the scheme of things it was a separate place marked off, proscribed and unwanted by the white people." The lesson imparted was absolutely clear. Whatever else Murray learned in school, she came to understand that her color marked her as inferior in the eyes of whites, regardless of how she conducted herself, regardless of how well she did in school, regardless of her social class.[6]

CONNECTIONS

When Darwin used the word *change*, some social Darwinists thought he meant *progress*. When Darwin described an organism as *different* from earlier ones, they assumed he meant the new organism was *better*. How do you account for such errors in reading? How do the times in which we live shape the ways we understand ideas? What other factors shape our thinking?

Sumner's goal was "a sound and natural social order." What do you think he meant? In what sense is a social order "natural"? Is there a natural way of organizing a society? What makes a society "sound"?

Each of us has a "universe of obligation"—a circle of individuals and groups toward whom we have obligations, to whom the rules of society apply, and whose injuries call for amends. Whom does Sumner consider "one of us"? Whom does he seem to exclude from citizenship? How does Sumner define his universe of obligation? Whom does Strong consider a fellow citizen? Whom does he seem to exclude? How does the Supreme Court define the nation's universe of obligation? Whom do the justices exclude?

What does Pauli Murray's story suggest about the consequences of the way many Americans defined their universe of obligation in the late 1800s and early 1900s? In what sense did the school Pauli Murray attended define her place in society? In what sense do schools in your community define your place in society? What other institutions in a community reflect how society regards particular individuals and groups?

In reflecting on the effect of a childhood in the Jim Crow South, Pauli Murray described herself as "not entirely free from the prevalent idea that I must prove myself." Yet by any standard, her accomplishments were impressive. At a time when few African Americans were able to even attend high school, she earned a college diploma (Hunter College in New York), a law degree (Howard University), and a Ph.D. (Yale University Law School). She became an attorney, a professor, a prize-winning author and poet, and an Episcopal priest. She was also an activist who challenged "Jim Crow" throughout her life. Long before the "sit-ins" and "freedom rides" that marked the Civil Rights Movement of the 1960s, Murray was arrested, jailed, and fined for refusing to sit in the segregated section of a Virginia bus. Find out more about Pauli Murray. What do her self-doubts suggest about the power of others to define not only one's place in society but also one's identity? To the importance of challenging that power?

---

1. Quoted in *In Search of Human Nature* by Carl Degler. Oxford University Press, 1991, p. 11.
2. *What Social Classes Owe to Each Other* by William Graham Sumner. New York, 1883, pp. 113–121.
3. *Our Country: Its Possible Future and Its Present Crisis* by Josiah Strong. Home Missionary Society, 1885, p. 168.
4. Ibid., p. 170.
5. *Before the Mayflower: A History of Black America* by Lerone Bennett, Jr. Penguin Books 1984, p. 256.
6. *Trouble in Mind: Black Southerners in the Age of Jim Crow* by Leon F. Litwack. Knopf, 1998, pp. 108–109.

Reading 2

By the late 1800s, the Industrial Revolution had changed not only how goods were made in the United States and much of Europe but also where they were made. More and more people were leaving the countryside for jobs in large urban centers, where they lived and worked among strangers. In such a society, it is all too easy to blame someone else for all that is new and disturbing in life. *They* are responsible for society's ills. Who are *they*? In the United States, *they* were African Americans, immigrants from Southern and Eastern Europe, Native Americans, and others who looked, spoke, or acted differently than *we* do.

Francis Galton, an English mathematician and Charles Darwin's cousin, offered an attractive solution to the threat *they* posed. He promised to "raise the present miserably low standard of the human race" by "breeding the best with the best." His theories were based on the idea that individuals are born with a "definite endowment" of qualities like "character, disposition, energy, intellect, or physical power"—qualities that in his view "go towards the making of civic worth."

Galton decided that natural selection does not work in human societies the way it does in nature, because people interfere with the process. As a result, the fittest do not always survive. So he set out to consciously "improve the race." He coined the word *eugenics* to describe efforts at "race betterment." It comes from a Greek word meaning "good in birth" or "noble in heredity." In 1883, Galton defined eugenics as "the science of improving stock, which is by no means confined to questions of judicious mating, but which . . . takes cognizance of all influences that tend in however remote a degree to give the more suitable races or strains of blood a better chance of prevailing speedily over the less suitable than they otherwise would have had."[1]

Galton was particularly concerned with the decline of genius in society. He believed that intelligence is an inherited trait and that the upper classes contain the most intelligent and accomplished people. He was therefore alarmed to discover that the poor had a higher birth rate. In 1904, Galton explained how eugenics might address that problem:

> Eugenics is the science which deals with all influences that improve and develop the inborn qualities of a race. But what is meant by improvement? We must leave morals as far as possible out of the discussion on account of the almost hopeless difficulties they raise as to whether a character as a whole is good or bad. The essentials of eugenics may, however, be easily defined. All would

> agree that it was better to be healthy than sick, vigorous than weak, well fitted than ill fitted for their part in life. In short, that it was better to be good rather than bad specimens of their kind, whatever that kind might be. There are a vast number of conflicting ideals, of alternative characters, of incompatible civilizations, which are wanted to give fullness and interest to life. The aim of eugenics is to represent each class or sect by its best specimens, causing them to contribute more than their proportion to the next generation; that done, to leave them to work out their common civilization in their own way.
>
> There are three stages to be passed through before eugenics can be widely practiced. First, it must be made familiar as an academic question, until its exact importance has been understood and accepted as a fact. Secondly, it must be recognized as a subject the practical development of which is in near prospect, and requires serious consideration. Thirdly, it must be introduced into the national conscience, like a new religion. It has, indeed, strong claims to become an orthodox religious tenet of the future, for eugenics cooperates with the workings of nature by ensuring that humanity shall be represented by the fittest races. What nature does blindly, slowly, and ruthlessly, man may do providently, quickly, and kindly. As it lies within his power, so it becomes his duty to work in that direction, just as it is his duty to be charitable to those in misfortune. The improvement of our stock seems one of the highest objects that can be reasonably attempted. We are ignorant of the ultimate destinies of humanity, but feel perfectly sure that it is as noble a work to raise its level as it would be disgraceful to abase it. I see no impossibility in eugenics becoming a religious dogma among mankind, but its details must first be worked out sedulously in the study. Over-zeal leading to hasty action would do harm by holding out expectations of a near golden age which would certainly be falsified and cause the science to be discredited. The first and main point is to secure the general intellectual acceptance of eugenics as a hopeful and most important study. Then let its principles work into the heart of the nation, which will gradually give practical effect to them in ways that we may not wholly foresee.[2]

Galton was not sure how to bring about these changes. Although he spent years studying heredity, by the time he died in 1911 he still had no idea how traits are passed from parent to child. In his research, however, Galton stumbled upon two discoveries that might have led another scientist to abandon eugenics. Neither fazed him. One was the result of a test he devised to measure intelligence. To his dismay, the poor did as well on the test as "the better elements in society." He concluded that the problem lay in the test rather than his theory.

His second discovery resulted from his efforts to track successive generations of pea plants. He found that, no matter how high the quality of the parent strains, some offspring were as good as the parent plant and some worse, but most were a little worse. This idea is known in statistics as "regression toward the mean" or middle. Galton suspected it was true for humans as well. If so, it would be impossible to improve the "race" through eugenics. Yet neither finding altered Galton's beliefs. He continued to insist that intelligence is linked to social class and that "the fittest" parents produce superior offspring.

CONNECTIONS

Compare and contrast Galton's definitions of eugenics. What are the key words in each definition? How are the two definitions alike? What differences are most striking? How do both definitions relate to Darwin's theory of natural selection?

Reread the first paragraph in Galton's 1904 description. What words or phrases stand out ("inborn qualities of the race," "better to be healthy than sick," etc.)? What does Galton say about eugenics? What does he imply? When Galton writes that the aim is for each "class or sect" to contribute its best elements to future generations, he is suggesting that all groups contribute to the future of humanity even though they are not equal. How do you think Galton expects each class to weed out its worst elements and find its appropriate place in society?

Galton insisted that the "best" people in a society are the "brightest." What is the power of that argument? How does it shape our society today?

What are the three stages Galton suggests as necessary to the success of eugenics? What is clear about each stage? What is vague? How do you account for the vagueness? Galton wanted eugenics to be accepted as an "orthodox religious tenet" and a scientific fact. Is it possible for an idea to be both a science and a religion? How does Galton seem to regard the relationship between science and society? The relationship between science and religion?

Why do you think Galton insisted that morals be left out of any discussion of eugenics as an "orthodox religious tenet"?

How is Galton's vision of a eugenic society similar to the "Masterpiece Society" described in Chapter 1? What differences seem most striking?

---

1. *Inquiries into the Human Faculty and Its Development* by Francis Galton. J.M. Dent and Sons, 1883, p. 24.

2. Reprinted by permission from *Nature,* May 26, 1904. Copyright 1904 Macmillan Magazines Ltd.

Reading 3

Francis Galton was aware that organisms within a species differ from one another. He also understood that each passes on characteristics to its offspring, but he did not know how offspring inherit traits from their parents. As a result, eugenics was little more than an interesting idea until scientists rediscovered Gregor Mendel's laws of heredity in the early 1900s.

Mendel did most of his research in the 1850s and 1860s, at about the time that Darwin was publicizing his theory of natural selection. Although Mendel also published his findings, few of his contemporaries paid attention to his work. No one knows why his studies were of so little interest at the time. Some historians believe that Darwin's theory of natural selection overshadowed every other idea in biology in the 1860s. Others observe that scientists at the time focused on ideas related to change and adaptation. Mendel's work, on the other hand, dealt with the way traits are passed on rather than with the way they change. Still other historians note that Mendel worked on a small scale at a time when most scientists were studying entire species. His work was also experimental and analytical at a time when many scientists were stressing description and speculation.

The son of peasants, Mendel studied at the universities of Olmütz and Vienna in the Austro-Hungarian Empire. In 1843, he entered a monastery in Brünn, partly because of his interest in research. The abbot had an experimental garden and was willing to support Mendel's work. Mendel began experimenting in 1857. Working with the green pea flower, he transferred pollen from a tall variety to the stigma of a short-stemmed variety. He sowed the resulting seeds to produce new plants whose characteristics offered insights into the relationship between parents and their offspring. People had been breeding animals and plants selectively for centuries even though they had no idea how inheritance worked. Most assumed that traits were passed through an organism's "bloodlines." Somehow, "blood" from both parents mingled together to create an offspring.

Based on his experiments with peas, Mendel disagreed. His experiments suggested that such traits as seed color and texture are inherited as discrete "particles." Either the offspring have a particular trait or they do not; there is no "mingling." To test his hypothesis, Mendel followed specific traits over many generations. He took groups of "pure line" smooth peas or wrinkled peas and fertilized them with their own pollen. ("Pure line" means that plants grown from these seeds, if self-fertilized, always duplicate the traits of the parent stock.) Pure-line smooth peas always produced more smooth peas.

Next, Mendel cross-fertilized varieties of peas. He found that when he crossed pure-line smooth with pure-line wrinkled, the first generation was always smooth. Usually, the second and third generations were smooth as well, but sometimes a plant with wrinkled peas would emerge. Over time, he noticed a pattern—there were about three smooth plants for every wrinkled one.

For nearly ten years Mendel combined multiple traits and carefully observed their appearance in successive generations. In 1865 in a paper presented to the Brünn Society for the Study of Natural Science, he described what became known as Mendel's Laws of Inheritance:

> **Principle of Dominance:** Each pea plant contains a set of hereditary particles (in 1909 they became known as genes). Alternate forms of the particles or genes are called alleles. If both alleles are the same, they are pure line or homozygous. If the alleles differ from one another, they are called hybrid or heterozygous. In the latter combination one trait always seems to be dominant and the other recessive. For example, a combination of smooth and wrinkled alleles will always yield smooth peas.
>
> **Principle of Segregation:** Mendel reasoned that the two matching alleles in each gene are segregated when reproductive cells (gametes) are formed. Therefore, each cell—the sperm or the egg—contains just one allele for a particular trait. When they come together in reproduction, the new seed contains an allele from each parent organism. Therefore, the reproductive cells, sperm and egg, contain one half of a gene pair, or one allele for each particular trait. When the organism reproduces, the new seed contains one allele from each gene pair in each parent organism.
>
> **Principle of Independent Assortment:** Different gene pairs defining different traits are passed on independently of each other in random combinations. Mendel crossed two hybrid pea plants with normal stature and smooth seeds (the dominant forms for these traits). The offspring included some with dwarf stature, some with wrinkled seeds, some with both, and most with neither. A plant that received one of these recessive traits was not more likely to receive the other. Mendel reasoned that the hereditary particles for different traits are not connected. (This later turned out to be true only in some cases.)

Mendel was lucky in his research. He experimented with a plant that was easy to grow and had a short generation time. Also a single gene affected each of the characteristics he studied. Because those characteristics are inherited separately, he could trace them individually. Without such luck, his experiments might have resulted only in confusion. Recent research indicates that most traits are not influenced by a single gene, but by several genes along with a variety of environmental and biological processes. Scientists also now know that a single gene may have multiple functions. Dominance is not always as clear-cut as it seemed to Mendel. Indeed, when he studied plants with complicated hereditary patterns, his predictions fell apart.

Mendel published his findings, but his work gathered dust in university libraries until 1900. That year, three scientists simultaneously discovered his writings. Each was working independently on problems involving hybrids. Within a very short time, they had introduced his ideas to dozens of other researchers. Little by little, these scientists enlarged Mendel's experiments to include more and more of the plant and animal kingdoms.

CONNECTIONS

What does Mendel's story suggest about the relationship between science and society? Why might some scientific theories be accepted immediately, while others are discounted or ignored for years?

To explore how Mendel's laws work, you may wish to try his experiment. It focuses on a single trait–color. As you work, keep in mind:

1. Every pea plant has two genes for determining color.
2. The green allele is dominant over the yellow one.
3. The genetic information within an organism is its genotype, (green/green; yellow/yellow; green/yellow).
4. The plant's external appearance (green peas or yellow peas) is its phenotype. (This was all Mendel himself could see.)

To illustrate the principle of dominance, place an equal number of green and yellow beads in a bag to symbolize the genes for color in Mendel's pea plants. Because each pea has two genes for color, reach into the bag and draw two beads at random. The two beads will determine the color of your pea plant.

- What is the genotype of your pea plant—GG, YY, or GY?
- What is the phenotype of your pea plant—green or yellow?
- How many combinations result in a green pea plant? In a yellow one?
- Repeat the process a few times. Which color is the more common?

To illustrate the principle of segregation, randomly select one bead and pair it with a bead from another student. You have just created a new "plant." One gene came from each parent plant. What is the genotype of your new pea plant? Its phenotype?

Reading 4

By the early 1900s, a number of scientists were trying to combine Gregor Mendel's research with Francis Galton's theory of "race improvement" so that they could tackle some of society's greatest problems. They viewed their work as a civic enterprise and claimed that eugenics would eventually reduce crime, end some diseases, and even boost human intelligence. It was a tempting vision—one that had particular appeal for middle class Americans in the early 1900s. It was a time when many marveled at the ability of science and technology to produce great wealth, create millions of new jobs, offer an ever-growing list of consumer goods, and open "life choices previously unimagined."

At the same time, many people were deeply troubled by the changes in their lives. As a result of their dis-ease, they were attracted to ideas that gave scientific meaning to the old rules and the old hierarchies. By 1915, eugenics had become a fad in the United States. Although the theory also attracted followers in England, Germany, Sweden, Denmark, Russia, Canada, and Brazil, the United States led the world in eugenic research in the first two decades of the 20th century. One of the most influential people in the American eugenics movement was Charles Davenport. While earning a Ph.D. in biology at Harvard University, he stumbled upon the writings of Francis Galton and other English eugenicists. Davenport was so taken with their ideas that he traveled to England to meet Galton. He returned home determined to incorporate eugenic principles into his own research.

Charles Davenport.

In 1904, Davenport persuaded the Carnegie Institution of Washington to provide the funding for the Station for Experimental Evolution at Cold Spring Harbor on Long Island in New York. He became its first director and oversaw early research into inheritance in both plants and animals. He hoped to combine Darwin's ideas on natural selection with Gregor Mendel's principles of heredity in controlled experiments.

By 1910, Davenport was prepared to go further. That year he established the Eugenics Record Office (ERO) at Cold Spring Harbor. There he and other researchers not only studied human heredity but also tried to demonstrate how social traits such as pauperism, criminality, and prostitution are inherited.

Davenport particularly wanted the ERO to educate the public about the importance of eugenic research in solving social problems. In 1911, he published a popular textbook, *Heredity in Relation to Eugenics,* for use in college and high school biology classes. The following excerpts illustrate some of Davenport's key assumptions and conclusions.

> Eugenics is the science of the improvement of the human race by better breeding or, as the late Sir Francis Galton expressed it:—"The science which deals with all influences that improve the inborn qualities of a race." The eugenical standpoint is that of the agriculturalist who, while recognizing the value of culture [environment], believes that permanent advance is to be made only by securing the best "blood." Man is an organism—an animal; and the laws of improvement of corn and race horses hold true for him also. Unless people accept this simple truth and let it influence marriage selection, human progress will cease. . . .
>
> There is no question that, taken as a whole, the hordes of Jews that are now coming to us from Russia and the extreme southeast of Europe, with their intense individualism and ideals of gain at the cost of any interest, represent the opposite extreme from the early English and the more recent Scandinavian immigration with their ideals of community life in the open country, advancement by the sweat of the brow, and the uprearing of families in the fear of God and the love of country. . . .
>
> Summarizing this review of recent conditions of immigration, it appears certain that, unless conditions change of themselves or are radically changed, the population of the United States will . . . rapidly become darker in pigmentation, smaller in stature, more mercurial, more attached to music and art, more given to crimes of larceny, kidnapping, assault, murder, rape, and sex-immorality and less given to burglary, drunkenness, and vagrancy than were the original English settlers. Since . . . there [are] relatively more foreign-born than native [in hospitals], it seems probable that under present conditions the ratio of insanity in the population will rapidly increase. . . .
>
> If increasing attention is paid to the selective elimination at our ports of entry of the actually undesirable (those with a germ plasm [genes] that has imbecile, epileptic, insane, criminalistic, alcoholic, and sexually immoral tendencies); if agents in Europe learn the family history of all applicants for naturalization; if the luring of the

credulous and suggestible by steamship agents abroad and especially in the south-east of Europe be reduced to its lowest limits, then we may expect to see our population not harmed by this mixture with a more mercurial people.[1]

CONNECTIONS

How does Davenport define eugenics? Compare his definition with Francis Galton's definition in Reading 2. On what points do the two writers agree? What differences seem most striking?

What is the effect of phrases such as "hordes of Jews," and "undesirables"? Who are the carriers of inferior "germ plasm"? Whom does Davenport consider "superior"? What traces of Camper's speculations about ideal types (Chapter 2) do you find in Davenport's work?

Davenport asserts that "human progress will cease" without eugenics. What does this suggest about the thousands of years of human history prior to 1900? He also asserts that Americans will become, on the average, shorter and darker than earlier generations. How does he seem to define human progress?

With whom do you think Davenport's book was particularly popular? Who do you think was most likely to oppose Davenport's ideas? How might these individuals and groups get heard in the early 1900s? or today?

---

1. *Heredity in Relation to Eugenics* by Charles Benedict Davenport. Henry Holt, 1911, pp. 1, 216, 219, 224.

Reading 5

How did Charles Davenport and other eugenicists prove that such traits as insanity, "criminalism," and sexually immoral tendencies were inherited? Were their investigations "good science"? Or, as Frederick Douglass once wrote, did they reason "from prejudice rather than from facts"? Charles Davenport's Eugenics Record Office (ERO) relied on family histories to track inherited traits from one generation to the next. To trace those histories, Davenport and his colleagues created pedigree charts or "family trees."

After a few weeks of training at the ERO, a field worker, often a college or graduate student, was sent to a poor house or an orphanage to observe behavior. There he or she would spot such traits as "shiftlessness," "criminalism," and "feeblemindedness." Another popular technique involved interviews with neighbors who offered their impressions of the person or family being studied. Davenport relied on his field workers for much of the data he used in his charts. Many of these field workers later held influential jobs at state mental hospitals, almshouses, and prisons.

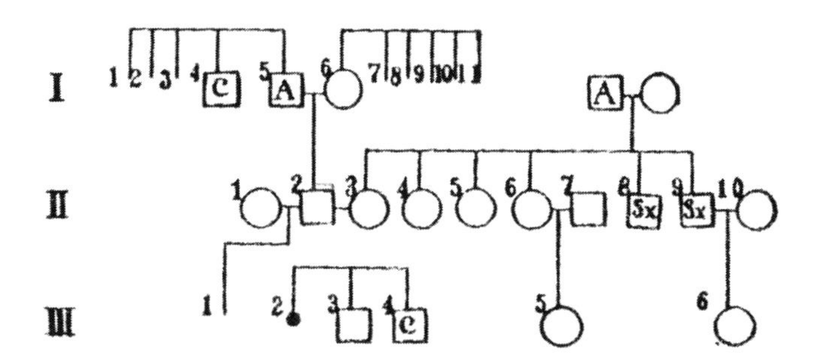

Pedigree chart used in *Heredity in Relation to Eugenics* by Charles Davenport.

Although no date is provided for the pedigree chart above, Davenport or one of his researchers probably gathered the information between 1887 and 1910. In Davenport's chart, the circles represent females and the squares stand for males. The Roman numerals indicate generations within a family. (I is first generation, II is second, and so on.) The other numbers refer to birth order among the children of a particular generation. The letters within the circles and squares designate a pronounced trait. "A" stands for "alcoholism," "C" for "criminality," and "Sx" for "sexual immorality." The narrative that accompanies the chart provides additional information about the fourth child, characterized by the letter "C", in the third generation of the family.

Figure 50, III, 4 is an eleven-year-old boy who began to steal at 3 years; at 4 set fire to a pantry resulting in an explosion that caused his mother's death; and at 8 set fire to a mattress. He is physically sound, able and well-informed, polite, gentlemanly and very smooth, but he is an inveterate thief and has a court record. His older brother, 14, has been full of deviltry, has stolen and set fires but is now settled down and is earning a living. Their father is an unusually fine, thoughtful, intelligent man, a grocer, for a time sang on the vaudeville stage; his mother, who died at 32, is said to have been a normal woman of excellent character. There is, however, a taint on both sides. The father's father was wild and drank when young and had a brother who was an inveterate thief. The mother's father was an alcoholic and when drunk mean and vicious. Some of the mother's brothers stole or were sexually immoral. . . .

The foregoing cases are samples of the scores that have been collected and serve as fair representations of the kind of blood that goes to the making of thousands of criminals in this country. It is just as sensible to imprison a person for feeble-mindedness or insanity as it is to imprison criminals belonging to such strains. The question whether a given person is a case for the penitentiary or the hospital is not primarily a legal question but one for a physician with the aid of a student of heredity and family histories.[1]

In creating such charts, Davenport assumed that a number of physical and behavioral traits are the result of a single recessive gene. He also believed that these traits were inherited in a simple Mendelian fashion. In some instances he was right. Huntington's chorea is the result of a single gene and so is color blindness. But for the most part, heredity is not nearly as simple as he believed. Even as Davenport claimed that wanderlust, pauperism, and criminality were inherited through the "unit characters" (genes) in one's blood, a number of scientists were undermining those claims.

Between 1905 and 1908, British geneticists Reginald Punnet and William Bateson, who coined the term *gene*, conducted experiments that suggested that the work of some genes modify the activity of others. Their research on sweet pea blossoms and the color of cock combs on roosters led to deeper insights into the ways genes act in combination with other genes to code for proteins and enzymes, which in turn influence such physical traits as skin color.

American biologist Thomas Hunt Morgan modified Mendel's ideas even further. Morgan conducted his experiments on the common fruit fly, *Drosophila melanogaster,* to find out how such physical traits as eye color and wing length are inherited. The fruit fly proved ideal for that kind of experimentation. It lives for just 14 days, is easy to breed, cheap to grow, and contains only 4 chromosome pairs. (Humans have 23 pairs.) Morgan discovered that genes are not randomly assorted, as Mendel had thought. Instead, genes that occur on the same chromosome are linked. Some traits are the result of a single gene as Davenport believed, but most are due to several genes working together. Morgan also found that the environment might alter the effects of particular genes on an organism. These discoveries greatly complicated ideas about heredity.

So did the research of Charles W. Stiles, a young scientist who worked for the U.S. Public Heath Service. He studied hookworm disease, which affected nearly 40 percent of all southerners in the late 1800s. A physician wrote that victims "become pale and anemic and complain of indigestion. In children, development, both physical and mental, is retarded and an infected child is dull and backward at school. In adults the symptoms vary with the intensity of the infection. A victim may feel weak, tire easily, and have shortness of breath. Also, infected persons may crave and eat unusual things such as paper, green fruit, chalk, clay and dirt—such persons are called 'dirt eaters.' Their muscles become weak, cause the abdomen to become prominent and enlarged, known as 'potbelly,' and the shoulder blades to stick out, 'angel wings.'"[2]

The disease was found primarily among poor people who lived in low-lying areas and lacked both shoes and sanitary facilities. With funding from the Rockefeller Foundation and the help of county health agencies, Charles Stiles organized a campaign to diagnose, treat, and eventually eliminate hookworm. Between 1909 and 1914, nearly 1.3 million people were examined for hookworm infection and 700,000 were treated. Stiles also initiated a campaign that stressed the importance of well-constructed out-houses and a good pair of shoes as ways of preventing the disease. The results were dramatic.

In Huntsville, Alabama, the local newspaper featured before-and-after photographs of a local family. The before photo showed a "tumbledown shack" where the family "lived in misery not knowing what was their trouble." The after photographs showed a family "so restored in health and vigor that they set to work to make enough money to better themselves in every way." One of the after photographs featured "the little white schoolhouse . . . where the children are now going to school to learn to read and write—things that were beyond the power and knowledge of their father and mother, their grandfathers and grandmothers, their great-grandfathers and great-grandmothers." The editors of the newspaper concluded the article by asking, "Is it any wonder that this family is doing what it can to prevent the further spread of the disease? Is it any wonder that the father has built a sanitary privy and is observing those simple rules of sanitation that if generally lived up to would completely banish hookworm disease from the country?"[3]

Even though Davenport was aware of the research of both Morgan and Stiles, he never addressed either man's research in his own scholarly works, textbooks, or lectures. Instead Davenport continued to assert that single genes are responsible for many physical and behavioral traits, including "feeblemindedness," "wanderlust," "pauperism," and "criminality."

CONNECTIONS

Notice the language used on page 78 to describe people featured on the pedigree chart (for example, "a normal woman of excellent character"). What judgments does the researcher make? What values do those judgments reflect? What evidence supports the claim that theft and arson are the results of a hereditary "taint"?

Review Mendel's laws of inheritance in Reading 3. How has Davenport applied Mendel's ideas in this family profile? Is heredity the only factor that may explain the boy's behavior patterns?

Why do you think Davenport ignored the findings of Morgan and Stiles? Why do you think their work did not make a difference in the way other scientists, politicians, and ordinary citizens viewed eugenics?

At first, Thomas Morgan, like nearly all biologists of the time, believed that the human condition could be improved by weeding out bad traits and enhancing positive ones. In time, he became one of the first biologists to criticize eugenics. In 1925, he wrote:

> Social reforms might, perhaps, more quickly and efficiently get at the root of part of the trouble, and until we know how much the environment is responsible for, I am inclined to think that the student of human heredity will do well to recommend more enlightenment in the social causes of deficiencies. . . . A little goodwill might seem more fitting in treating these complicated questions than the attitude adopted by some of the modern race propagandists.[4]

What questions might Morgan raise about Davenport's pedigree chart? What questions do you have about the chart? What additional information would you like to have?

---

1. *Heredity in Relation to Eugenics* by Charles B. Davenport. Henry Holt, 1911.
2. *The Germ of Laziness: Rockefeller Philanthropy and Public Health in the New South* by John Ettling. Harvard University Press, 1981, p. 4.
3. Ibid., p. 149.
4. *Evolution and Genetics*, 2d ed., by Thomas Hunt Morgan. Princeton University Press, 1925, p. 201.

Reading 6

Charles Davenport insisted that intelligence and other traits are transmitted in a Mendelian fashion despite scientific research to the contrary. To popularize that stand, he and a number of other eugenicists authored books that traced the history of a single family to prove that "feeblemindedness" is a hereditary trait. In 1912, Henry H. Goddard published the most popular of these studies. Entitled *The Kallikak Family: A Study in the Heredity of Feeblemindedness*, the book went through twelve editions between 1912 and 1939 and was widely quoted in not only academic journals and scholarly tomes but also popular magazines and high school textbooks. There was even talk of turning the book into a Broadway play.

Goddard's book was unusual in that it compared two branches of the same family—one respectable and the other "a race of defective degenerates." Although the family was real, the name is an alias that Goddard created by combining the Greek words for "beautiful" (kalos) and "bad" (kakos). Originally Goddard, who directed a laboratory for the study of mental deficiency at the Vineland Training School for Feeble-minded Boys and Girls in New Jersey, planned to focus his research on the direct ancestors of an inmate at the school—a young woman he called "Deborah Kallikak." Through interviews with her living relatives, Goddard's chief researcher, Elizabeth Kite, traced Deborah's family tree back to a great-great-grandfather, "Martin Kallikak." Kite also located another branch of the family with the same last name but with a markedly different reputation. At Goddard's request, she studied the history of that family as well. Based on her efforts, Goddard concluded:

> A young man of good family becomes through two different women the ancestor of two lines of descendants—the one characterized by thoroughly good, respectable, normal citizenship, with almost no exception; the other being equally characterized by mental defect in every generation. . . . We find on the good side of the family prominent people in all walks of life. . . . On the bad side we find paupers, criminals, prostitutes, drunkards, and examples of all forms of social pest with which modern society is burdened.
>
> From this we conclude that feeble-mindedness is largely responsible for these social sores. Feeble-mindedness is hereditary and transmitted as surely as any other character. We cannot successfully cope with those conditions until we recognize feeble-mindedness and its hereditary nature, recognize it early, and take care of it.[1]

How were Goddard and Kite able to assess the character and intelligence of people who had died over a hundred years earlier? Goddard told readers that "after some experience, the field worker becomes expert in inferring the condition of those persons who are not seen from the similarity of the language used in describing them to that used in describing persons whom she has seen." Goddard even included some of Kite's reports in his book to "show something of her method, and enable the reader to judge of the reliability of the data." In one of those reports Kite notes that a 12-year-old girl should have been in school, "but when one saw her face, one realized that it made no difference. She was pretty . . . but there was no mind there." She describes three children as having the "unmistakable look of the feeble-minded."[2]

Only a few scholars openly criticized the methods used by Goddard and other eugenicists. One of the most outspoken was Abraham Meyerson, a professor of neurology at Tufts University.

> I have had charge of a clinic where alleged feebleminded persons were brought every day, and I see in my practice and hospital work murderers, thieves, sex offenders, failures, etc. Many of these are brought to me by social workers, keen intelligent women, who are in grave doubt as to the mental condition of their charges after months of daily relationships, after intimate knowledge, and prolonged effort to understand. . . . And I have to say of myself, with due humility, that I have had to reverse my first impressions many and many a time.
>
> Judge how superior the field workers trained by Dr. Goddard were! Not only does "their first glance" tell them that a person is feebleminded, but they even know, without the faintest misgiving, that a "feebleminded girl" living over a hundred years before in a primitive community is feeble-minded. They know this, and Dr. Goddard, acting on this superior female intuition, founds an important theory of feeblemindedness, and draws sweeping generalizations, with a fine moral undertone, from their work. Now I am frank to say that the matter is an unexplained miracle to me. How can anyone know anything definite about a nameless girl, living five generations, before, whom no one has ever seen?[3]

Despite such criticism, studies like Goddard's remained popular with scholars and ordinary citizens. That popularity had real consequences for "Deborah Kallikak" and other Americans who were labeled as "feebleminded." Many, including Kallikak, spent much of their lives in hospitals, "training schools," and other institutions. Late in her life, Deborah had an opportunity to leave the school. Although as a young girl, she tried to escape from Vineland, she now chose to stay, because she required constant medical attention and no longer had ties to anyone in the outside world. When she died in 1978 at the age of 79, she was described as a "wonderful lady" who engaged in many community activities. She said of herself, "I guess after all I am where I belong. . . . I don't like this feeble-minded part, but anyhow I am not idiotic like some of the poor things around here. . . . Here everybody who is anybody, knows all about me and what I can do. With the wonderful friends that I got and the work I like so much, this place is my home."

CONNECTIONS

In the mid-1800s, Frederick Douglass wrote, "It is the province of prejudice to blind; and scientific writers, not less than others, write to please, as well as to instruct, and even unconsciously to themselve, (sometimes,) sacrifice what is true to what is popular." To what extent did prejudice blind Goddard?

How might Goddard and Kite have answered the questions Meyerson raises? Why do you think neither seemed to doubt the value of "first impressions"?

Despite Meyerson's criticisms, many scholars continued to cite Goddard's research long after other scientific research raised important questions about his methodology and scientific assumptions. How do you account for their support?

In 1949, a researcher said of Goddard's study, "The assistants whom he employed to secure his genealogical records had relatively little training but were fired with Goddard's enthusiasm. That they may have sometimes tended to find mental defect where mental defect was to be expected was perhaps inevitable under the circumstances, but no one can doubt the sincerity of their attempts to get the facts." What are the qualities of a good researcher? If you had to choose one of those qualities as more important than any other, what would it be? Where do such qualities as "enthusiasm" and "sincerity" rank on your list?

When Deborah Kallikak was admitted to Vineland, she was eight years old. Her admissions report claims she was able to dress herself, recognize a few letters, understand commands, and sew. She was also described as "obstinate and destructive" and "not very obedient." Over the years, Deborah's teachers at Vineland described her as:

> —Learning a new occupation quickly, but requires a half-hour or twenty-four repetitions to learn four lines.
> —Retaining well what she has once learned. Needs close supervision.
> —Bold towards strangers, kind towards animals.

—Able to run an electric sewing machine, cook, and do practically everything about the house.
—Having no notable defect. She is quick and observing, has a good memory, writes fairly, does excellent work in woodcarving and kindergarten (where she is an assistant), is excellent in imitation.
—Doing fine basketry and gardening. Spelling is poor; music is excellent; sewing excellent; excellent in entertainment work. Very fond of children and good in caring for them.
—Having a good sense of order and cleanliness. Sometimes very stubborn and obstinate.
—Not always truthful; has been known to steal, although does not have a reputation for this.
—Proud of her clothes. Likes pretty dresses and likes to help in other cottages, even to temporarily taking charge of the group.[4]

As an adolescent, Deborah often got in trouble with authorities. One report noted that "her skill with woodworking tools made it possible to alter her window screen and this fact, together with a moonlit campus and a convenient lover, set the stage for a romantic interlude. This had not progressed far when it was fortunately discovered. The young man was kindly dismissed by a lenient justice of the peace but Deborah per force remained in our custody." A few years later, the staff tried to place her in a nearby community only to learn that she had once again found a boyfriend. She was promptly returned to Vineland "in sack cloth and ashes." Deborah noted, "It isn't as if I'd done anything wrong. It was only nature."

To Henry Goddard, Deborah was a "typical illustration of the mentality of the high-grade, evil-minded, the moron, the delinquent, the kind of girl or woman who fills our reformatories." How do her teachers regard Deborah? How would you characterize her? Was she "where she belonged"? What voice did she have? Who spoke for her?

According to Goddard, the moral of "Deborah Kallikak's" story is that feeblemindedness is hereditary and dangerous. What other morals might one draw from her story? What does it suggest about the power of labels? The role of the environment in shaping identity? The power of education to transform an individual?

---

1. *The Kallikak Family: A Study in the Heredity of Feeblemindedness* by H.H. Goddard. Macmillan, 1912, pp. 115–116.
2. Ibid., p. 76.
3. Quoted in *The Legacy of Malthus* by Allan Chase. Alfred A. Knopf, 1977, pp. 122-123.
4. *A History of Mental Retardation* by R.C. Scheerenberger. Paul H. Brookes Publishing Co., 1983, p. 150.

Reading 7

In the early 1900s, most scientists in the United States viewed humanity as the sum of inherited traits and were convinced that some races were superior to others. Therefore many supported the segregation and isolation of the "feeble-minded" and the mentally ill. They also supported laws that separated black and white Americans, kept the Chinese from entering the nation as immigrants, and relegated Native Americans to "reservations." Only a few raised troubling questions about race and heredity. Among those scientists was the nation's leading anthropologist, Franz Boas.

Boas, a German immigrant with degrees in physics and geography, settled in the United States in the late 1800s. His career in anthropology began when he joined an expedition to the Cumberland Sound in Baffinland, Greenland, in 1883. He wrote of the experience:

> [It] was with feelings of sorrow and regret that I parted from my Arctic friends. I had seen that they enjoyed life, and a hard life, as we do; that nature is also beautiful to them; that feelings of friendship also root in the Eskimo heart; that although the character of their life is so rude as compared to civilized life, the Eskimo is a man as we are; that his feelings, his virtues, and his shortcomings are based in human nature, like ours.[1]

Boas's stay in Greenland led him to question his own assumptions about race and the meanings he and others attached to human differences. In 1894, he gave his first scholarly address on the topic. In it, he argued that "historical events appear to have been much more potent in leading races to civilization than [inherited ability], and it follows that achievements of races do not warrant us to assume that one race is more gifted than the other."

Boas believed that "the physiological and psychological state of an organism at a certain moment is a function of its whole history." He responded to those who asked why some groups seemed unable to absorb Western civilization with the suggestion that they look to history, experience, and circumstances, rather than race, for answers.

As a professor, Boas challenged his students to put aside their prejudices in studying other cultures, or ways of life. As a curator at the American Museum of Natural History, he insisted that all of the artifacts belonging to a particular group be placed together so visitors could see how they related to one another.

He argued against arranging artifacts in ways that suggested that some "races" were superior to others. As a scholar, he demanded that those who looked to biology or race to explain human differences prove their claims. And he insisted over and over again that mental differences between the races "have not been proved yet."

As a citizen, Boas felt strongly about equal opportunity. He came to the United States because of the discrimination he experienced as a Jew in Germany. As an American, he was particularly troubled by the plight of African Americans. In 1905, W. E. B. DuBois, the leading African American social scientist of his day, invited Boas to speak at Atlanta University, an all-black college. In *Black Folk Then and Now*, DuBois recalled the visit:

> Franz Boas came to Atlanta University where I was teaching history . . . and said to the graduating class: You need not be ashamed of your African past; and then he recounted the history of black kingdoms south of the Sahara for a thousand years. I was too astonished to speak. All of this I had never heard and I came then and afterwards to realize how the silence and neglect of science can let truth utterly disappear or even be unconsciously distorted.

A few months later, Boas wrote a letter to Andrew Carnegie asking for his help in establishing an "African Institute." The letter states in part:

> The increasing antagonism between the white and the black races is not only a matter of concern from a humanitarian point of view, but entails serious dangers to the Commonwealth. Notwithstanding all that has been written and said on the subject of racial ability or inability of the Negro, a dispassionate investigation of the data at hand shows neither that his inability has ever been demonstrated, nor that it has been possible to show that the inferiority of the Negro in America is entirely due to social rather than to racial causes. . . .
>
> It seems plausible that the whole attitude of our people in regard to the Negro might be materially modified if we had a better knowledge of what the Negro has really done and accomplished in his own native country.
>
> It would seem that any endeavor of this kind should be connected with thorough studies of the conditions of the American Negro on such scientific basis that the results could not be challenged. The endless repetition of remarks on the inferiority of the Negro physique, of the early arrest of the development of Negro children, of the

> tendency in the mulatto to inherit all the bad traits of both parental races, seems almost ineradicable, and in the present state of our knowledge can just as little be repudiated as supported by definite evidence.
>
> There seems to be another reason which would make it highly desirable to disseminate knowledge of the achievements of African culture, particularly among the Negroes. In vast portions of our country there is a strong feeling of despondency among the best classes of the Negro, due to the economic, mental, and moral inferiority of the race in America, and the knowledge of the strength of their parental race in their native surroundings must have a wholesome and highly stimulating effect. I have noticed this effect myself in addressing audiences of Southern Negroes, to whom the facts were a complete revelation.
>
> Considering that the future of millions of people is concerned, I believe that no energy should be spared to make the relations of the two races more wholesome, and to decide by unprejudiced scientific investigation what policy should be pursued. I should be inclined to think that an institution which might be called "African Institute" could contribute materially to the solution of these problems. Its purpose ought to be the presentation to the public, by means of exhibits and by means of publications, of the best products of African civilization. This should be accompanied by a scientific study of this civilization—one of the most important means of creating a group of men who will intelligently present the subject.
>
> A second division of such an institution should be devoted to the study of the anatomy of the Negro. The investigations of such a division would be necessarily technical, but they would have the most important bearing upon the question of general policy to be pursued in regard to the Negro. . . .
>
> A third division of such an institution should be devoted to statistical inquiries of the Negro race in this country; and here, also, I believe the most useful work could be done.[2]

Boas was unable to secure funding for his project from any of Carnegie's many foundations. Yet one of those foundations supported Charles Davenport's Eugenics Record Office. It also gave large sums of money to Booker T. Washington's Tuskegee Institute, primarily to provide African Americans with vocational training.

CONNECTIONS

What does W. E. B. DuBois mean when he recalls that after Boas's commencement address, he came "to realize how the silence and neglect of science can let truth utterly disappear or even be unconsciously distorted." What is he suggesting about the power of silence? What does he believe is necessary to keep truth alive?

In the late 1800s, few people paid much attention to Franz Boas's ideas about race. Yet popular magazines and newspapers carried article after article boasting of the superiority of the "Anglo-Saxon" and the inferiority of other races. They also printed articles by "race scientists," eugenicists, and anthropologists who believed that race explained all human differences. Why do you think some ideas become popular very quickly, while others are viewed with suspicion, even fear? What ideas are more difficult to believe? What ideas are most disturbing? Threatening?

Boas actively encouraged African Americans to become anthropologists. One of his students was Zora Neale Hurston. Born in Eatonville, Florida, Hurston was the first African American woman to graduate from Barnard College. As an anthropologist, she traveled through the South tracing the folklore of African Americans. Her research offered insights into a forgotten history and encouraged the study of folklore worldwide. How does Hurston's work deepen our understanding of the importance Boas placed on training anthropologists of all races and ethnicities?

The word *culture* is often defined as a way of life. It shapes not only how people live, work, and play but also their attitudes, values, and beliefs. We view the world through the lens of our culture. What does Boas suggest about how we can learn to view the world through another cultural lens? Why do you think he believed it was important to do so? Boas came to the United States from Germany. How do you think his experiences as an immigrant may have shaped the value he placed on looking at the world through multiple perspectives?

---

1. *From Savage to Negro* by Lee D. Baker. University of California Press, 1998, pp. 36.
2. *The Franz Boas Reader*, edited by George W. Stocking, Jr. University of Chicago Press, 1974.

4. In an Age of "Progress"

We are all of us immigrants in the industrial world, and we have no authority to lean upon. We are an uprooted people, newly arrived and nouveau riche.

Walter Lippmann

Chapter 3 described the growth of eugenics, a branch of scientific inquiry developed by Francis Galton, an English mathematician. He based the new science on the idea that individuals are born with a "definite endowment" of qualities like "character, disposition, energy, intellect, or physical power"—qualities that in his view "go towards the making of civic worth." Eugenics therefore promised to "raise the present miserably low standard of the human race" by "breeding the best with the best."

Chapter 4 considers how eugenics was related to other aspects of American life at the turn of the 20th century. Many of the readings place the movement in an historical context by focusing on some of the changes that transformed American life in the late 1800s and early 1900s. The Industrial Revolution had swept away familiar ways of working and living, altered social expectations, and redefined the relationship between citizens and their government. In a book of reminiscences entitled *The Age of Confidence*, editor Henry S. Canby wrote of his own responses to those changes and those of other white middle-class Americans in the late 1800s:

> We had been trained to fit into certainties, educated to suppose that Mr. [Andrew] Carnegie's steel mills, Sunday observance, the banking system, the Republican party, the benefits of Latin, algebra, and good handwriting . . . were parts of one quite comprehensible plan. . . . Yet whispering at the back of the new liberal mind was always a question which became more insistent as the years went on. The community in which we had been brought up and the education ground into us were ordered, self-contained, comprehensible, while this new society was incoherent, without fixed aim, and without even a pretense of homogeneity. We were like pond fish who had been flooded into a river.[1]

Americans like Canby were ambivalent about change. Their pride in the nation's scientific advances and technological innovations was tempered by their discomfort with social and economic transformations. A number of them looked back at the world they had known as children with a deep sense of loss. Each year fewer Americans made their home on farms or in small towns where people knew their neighbors. More and more now lived and worked among strangers in huge metropolitan areas. By 1900 New York City was home to over 4 million people. Chicago had a population of over 1.7 million and Philadelphia 1.4 million. Some smaller cities were doubling and tripling in population in the course of a decade.

To a growing number of middle-class white Americans, the city represented all that was new and disturbing in their world. In *Our Country*, one of the most popular books of the era, author Josiah Strong, a Protestant minister, described the "seven perils" that he claimed threatened the nation. The first six were Catholicism, "Mormonism," intemperance, socialism, wealth, and immigration. The seventh peril was the city itself—the base for the "alien army that invaded the nation," "an army twice as vast as the estimated numbers of Goths and Vandals that swept over Southern Europe and overwhelmed Rome."[2]

Beginning in the late 1800s, a number of middle-class white Americans set out to save "civilization" from the "perils" Strong and others described. Known as "progressives," these Americans tried to make their chaotic world more rational by tackling problems caused by rapid industrialization, migration, immigration, and urbanization. Unlike social Darwinists who believed in the survival of the "fittest," progressives believed they had a duty to intervene in society, a responsibility to help the less fortunate become as "fit" as possible. These Americans placed their faith in education and legislation. They were not an organized group, although they shared similar views on the dangers of child labor, overcrowded neighborhoods, and unsanitary living conditions. Their numbers included Democrats, Republicans, and independents. Although most were middle-class white Americans, on some issues they had the support of labor union leaders, immigrants, African Americans, and even wealthy industrialists.

---

1. Quoted in *Life in Twentieth Century America* by John W. Dodds. G.P. Putnam's Sons, 1965, 1972, p. 52.
2. Quoted in *The Free and the Unfree* by Peter N. Carroll and David W. Noble. Penguin Books, 1977, 1988, p. 240.

Reading 1

Many Americans at the turn of the 20th century viewed the changes that had taken place in their lifetimes with pride and amazement. In 1889, author Samuel Clemens, better known as Mark Twain, expressed those feelings in a letter to congratulate poet Walt Whitman on his 70th birthday:

> You have lived just the seventy years which are greatest in the world's history and richest in benefit and advancement to its peoples. These seventy years have done more to widen the interval between man and the other animals than was accomplished by any of the five centuries which preceded them.
>
> What great births have you witnessed! The steam press, the steamship, the steelship, the railroad, the perfect cotton gin, the telegraph, the phonograph, the photogravure, the electrotype, the gaslight, the electric light, the sewing machine and the amazing infinitely varied and innumerable products of coal tar; those latest and strangest marvels of a marvelous age. And you have seen even greater births than these; for you have seen the application of anesthesia to surgery-practice, whereby the ancient dominion of pain, which began with the first created life, came to an end on this earth forever; you have seen the slave set free, you have seen monarchy banished from France and reduced in England to a machine which makes an imposing show of diligence and attention to business, but isn't connected with the works. Yes you have indeed seen much—but tarry for a while, for the greatest is yet to come. Wait thirty years, and then look out over the earth! You shall see marvels upon marvels added to those whose nativity you have witnessed; and conspicuous above them you shall see their formidable Result—man at almost his full stature at last!—and still growing, visibly growing while you look. Wait till you see that great figure appear; and catch the far glint of the sun upon his banner; then you may depart satisfied, as knowing you have seen him for whom the earth was made, and that he will proclaim that human wheat is more than human [seeds], and proceed to organize human values on that basis.[1]

Had Whitman lived until the turn of the century, he would have witnessed many more of the benefits of "a marvelous age." Historian John Milton Cooper, Jr. writes that by 1900:

Not only had the United States grown to continental size, but its population had swelled to seventy-six million, spread from coast to coast in forty-five states, and concentrated in thirty-eight cities of more than one hundred thousand people. In 1900, no aspect of American life was more striking that this rapid, fantastic growth. The ballooning numbers of people sprang in part from a high, but now declining, annual birth rate: 32.3 live births per thousand of population (down from 55 in 1800 and 43.3 in 1850.) Greater growth resulted from lowered infant mortality and lengthened life span, which had reduced the annual death rate to 16.5 per thousand, the lowest in the world. But by far the greatest numbers of new Americans came with the waves of immigration from overseas. Nearly 425,000 Europeans arrived . . . in 1900 alone.

Americans were proud of the drawing power of their political and religious freedoms, which had long since made them a "nation of immigrants." From the beginning of the nineteenth century, European migration to the United States had steadily mounted and had become more diverse than in the colonial period, when most settlers had been English and Scottish Protestants. Starting in the 1840s, thousands of Irish immigrants, most of whom were Roman Catholics, as well as Germans of various religious persuasions, flocked across the ocean. After the Civil War, the sources of European immigration broadened still further to encompass growing numbers from Scandinavia, Italy, Greece, and Eastern Europe. . . . In 1900, the rate of immigration was still accelerating. During the first decade of the twentieth century, over eight million more immigrants would come to the United States—the largest number in any decade before or since. These newest arrivals would account for more than 10 percent of the entire American population.

Size, population, wealth—each marked how far the United States had come in such a short time from its raw, humble beginnings. Only two countries, Russia and Canada, occupied larger land areas. Among the Western nations—those with predominately European ethnic origins, languages, and cultures—only Russia had a larger population. No country anywhere enjoyed so large and dynamic an economy. American commerce, transportation, industry, and agriculture were wonders of the world. By almost any measure of economic performance, the United States excelled. Steel production in 1900 amounted to over ten million tons, more than a third higher than Germany's, the closest competitor. Railroad trackage stretched to 167,000 miles, or one-third of the world's total. Per-capita income was estimated at $569, far above the nearest rival, Britain. Literacy rates stood at nearly 90 percent of the populace. The country had over 2,200 newspapers and nearly one thousand colleges and universities, with a combined student body of nearly 240,000. School enrollment amounted to over sixteen million pupils—the world's largest in both numbers and percentage of the population. Of these students, nearly one hundred thousand would graduate from secondary schools in 1900, also ahead of every other nation in numbers and percentages, and nearly double the total in 1890.[2]

CONNECTIONS

Some people define the word *progress* as "growth" or "movement," while others view it as "a step forward" or a "ladder reaching upward." How does Mark Twain define the word? What achievements does he regard as central to progress? How do you define *progress*? How does the way one defines the term shape an understanding of the world?

Scientist Jacob Bronowski created "The Ascent of Man," a television series and a book on the history of humankind. He explained his use of the word *ascent*: "Man ascends by discovering the fullness of his own gifts (his talents or faculties) and what he creates on the way are monuments to the stages in his understanding of nature and self."[3] How is his view of *ascent* similar to Twain's view of *progress*? To those expressed by people like Samuel Morton (Chapter 2) and Charles Davenport (Chapter 3)? How do these views of progress differ? Which view is closest to your own definition of the term?

How does the word *progress* apply to individuals? What does it mean to regard yourself and others as those "for whom earth was made"? How does that view shape the way Twain ranks humankind in relation to other animals? How are his efforts to arrange the natural world similar to those of Johann Blumenbach or Petrus Camper (Chapter 2)? How do you think someone like Charles Davenport would respond to Twain's view of the nation's future? On what might they agree? On what points might there be debate?

Cooper writes, "During the first decade of the twentieth century, over eight million more immigrants would come to the United States—the largest number in any decade before or since. These newest arrivals would account for more than 10 percent of the entire American population." How do you think these newcomers may have contributed to the dis-ease Henry Canby describes in the

Introduction to Chapter 4? What other signs can you find in Cooper's account that might explain the dis-ease experienced by Americans like Canby? Their sense of loss?

Interview someone who has lived 70 years or more to find out what changes have taken place in the world in his or her lifetime. How might Twain have described those changes? Which might he regard as "marvels of a marvelous age"? If he were alive today, how might he have revised or expanded his assessment of the marvels of his own age? His assessment of the future of humankind?

---

1. Quoted in *Letters of a Nation* edited by Andrew Carroll. Kodansha America, Inc., 1997, pp. 396-397.
2. *Pivotal Decades: The United States, 1900-1920* by John Milton Cooper, Jr. W.W. Norton & Co., 1990, pp. 1-3.
3. *The Ascent of Man* by Jacob Bronowski. Little Brown, & Co., 1973, p. 24.

The End of the Frontier

Reading 2

In 1890, the Census Bureau announced that the nation had become so settled that it was no longer possible to draw a line on a map of the United States to indicate the nation's frontier. Historian Frederick Jackson Turner saw the announcement as the end of an era. In speeches and essays, he maintained that with the closing of the frontier, something distinctive and even precious in American life had been lost. In 1926, journalist Mark Sullivan mourned that loss in *Our Times*, a history of the early 1900s. He argued that at the turn of the 20th century, "the average American in great numbers had the feeling he was being 'put upon' by something he couldn't quite see or get his fingers on; that somebody was 'riding' him; that some force or other was 'crowding' him." Sullivan explained:

> Vaguely he felt that his freedom of action, his opportunity to do as he pleased, was being frustrated in ways mysterious in their origin and operation, and in their effects most uncomfortable; that his economic freedom, as well as his freedom of action, and his capacity to direct his political liberty toward results he desired, was being circumscribed in a tightening ring, the drawing-strings of which, he felt sure, were being pulled by the hands of some invisible power which he ardently desired to see and get at, but could not. This unseen enemy he tried to personify. He called it the Invisible Government, the Money Interests, the Gold Bugs, Wall Street, the Trusts. During the first [William Jennings] Bryan campaign [for President in 1896], the spokesmen of the West spoke of the businessmen of the East, collectively, as "the enemy."
>
> That mood was the source of most of the social and political movements of the years succeeding 1900. . . .
>
> The principal cause of the loss by the average American of a degree of economic freedom he had been accustomed to enjoy since the first settlement of the country was the practical coming to an end of the supply of free, or substantially free, virgin land. . . . During the 1890s occurred the last important one of these openings of Indian reservations to settlement, which were the principal means by which the Federal Government gave opportunity to landless men to acquire farms at small cost. That marked the end of that gloriously prodigal period . . . during which a man with a family of sons need give little concern to their future, knowing that when the urge of manhood

came, they could go out and acquire a farm by little more than the process of "squatting" upon it. The time had come to an end when a man of independent spirit, feeling distaste for going to work as any one's hired man in a factory or elsewhere, could go West, settle upon a quarter-section of public land, and in course of time possess himself of it without being called on to pay more than a nominal sum. The average American, who had been able to look out on a far horizon of seemingly limitless land, now saw that horizon close in around him in the shape of the economic walls of a different sort of industrial and economic organization, walls which, to be sure, could be climbed; but which called for climbing. . . .

The end of free land was the largest one of those causes which, in the years preceding 1900, gave rise to a prevailing mood of repression, of discomfort, sullenly silent or angrily vocal. . . . It took time to pass from an easy-going assumption that our land, our forests, all our natural resources were unlimited, to uncomfortable consciousness that they were not. The average American, more readily visualizing a personified cause for his discomfort, dwelt more upon causes that proceeded from persons, or organizations of persons—corporations, "trusts," or what-not. There were such causes. But they were minor compared to the ending of the supply of free land.

. . . . In 1900, many men could remember when they could take their rifles, go out among the buffalo-herds, and get as much meat as they wanted, without . . . hindrance. To men with that memory, regulations, hunters' licenses, were irksome. This is a small illustration of what happened in many fields. The frontiersman had hardly ever encountered law or regulation. With increase of population came limits on liberty, "verbotens," "forbidden by law," "no trespassing." Later, with machinery, came another variety of regulation. In the days of the horse-drawn vehicle, "keep to the right" was about the only traffic code. With the coming of the automobile, stringent traffic rules came into being.[1]

CONNECTIONS

Whom does Mark Sullivan regard as the "average American"? How does he describe the mood of that "average American" at the turn of the 20th century? To what does he attribute that mood? Why does he see it as the "source of most of the social and political movements in the years succeeding 1900"? As you continue reading, look for evidence that supports or challenges Sullivan's views.

Each of us has a "universe of obligation"—a circle of individuals and groups toward whom we feel obligations, to whom the rules of society apply, and whose injuries call for amends. Whom does Sullivan consider "one of us"? Who lies beyond his universe of obligation?

Sullivan focuses on life at the turn of the 20th century. In looking at that same period, anthropologist Lee Baker expresses concerns about the role of the "average American" in "the violent chaos that erupted at the massacre at Wounded Knee in 1890, race and labor riots in 1892, terrorizing lynch mobs, and reports that African Americans composed the most criminal element in society." To what extent is there a connection between "feeling put upon" and outbreaks of violence? Historians have noted that in times of uncertainty, it is all too easy to blame someone else for all that is new and disturbing. Whom does Sullivan's "average American" blame for his troubles? What do your answers suggest about the conditions that seem to encourage intolerance? What conditions then might foster tolerance? Find examples in current events.

Look up the words *squatting* or *squatter* in a dictionary. What do the definitions suggest about the way some Americans acquired "free land"?

How does Sullivan define the word *liberty*? What relationship does he see between individual liberties and the law? How do you define that relationship?

Many historians today disagree with the views expressed by Turner and Sullivan. In the book *Into the West*, historian Walter Nugent writes that by 1890 "Native American armed resistance had collapsed after four hundred years of European pressure. That, not the frontier, was what really ended in 1890." What point is Nugent making about the settlement of the West and the role of Native Americans in the process? Find out more about the frontier in American history. To what extent is the picture Sullivan paints reality? To what extent is it a myth? It has been said that what people believe is true often has more power than truth itself. How does the popular view of the settlement of the West support that idea?

In 1776, soon after the American Revolution began, each of England's 13 former colonies wrote a constitution that gave the right to vote to "free men" who owned property. By the mid-1800s, most states had revised their constitutions to allow all "free white men" to vote. What does Sullivan suggest about the links between land ownership and citizenship? Why do you think the Americans he describes felt that they had a right to the land?

---

1. *Our Times: The United States 1900-1925* by Mark Sullivan. Charles Scribner's Sons, 1926, pp. 137-138, 141-143, 144-149.

Reading 3

In the late 1800s and early 1900s, expositions and fairs were a way of educating people not only about their nation and its place in the world but also about their own place in American society. In 1893, over 27 million people attended the World's Columbian Exposition—an exposition that used architecture, artifacts, and "living exhibits" to celebrate "American progress." Held in Chicago to mark the 400th anniversary of Christopher Columbus's voyages to the Americas, it attracted over 13 million Americans—about one of every five people in the nation. The fair was designed to prove that "the wonderful progress of the United States, as well as the character of the people," is the result of natural selection. Many of the exhibits illustrated "the steps of progress of civilization and its arts in successive centuries, and in all lands up to the present time." The aim was "to teach a lesson; to show the advancement of evolution of man." That lesson was rooted in social Darwinism—the idea that competition rewards "the strong" (Chapter 3).

A view of the "White City," as the World's Columbian Exposition was known.

That kind of patriotism appealed to many Americans, including Francis J. Bellamy, an editor of the popular children's magazine *Youth's Companion.* At his urging, Congress made October 12, 1892, a national holiday. On that day children gathered at schools and churches to celebrate Columbus's achievements and the fair by reciting a "Pledge of Allegiance" that Bellamy wrote for the occasion: "I pledge allegiance to my Flag and the Republic for which it stands; one nation indivisible with liberty and justice for all." At the exposition, hundreds of schoolgirls dressed in red, white, and blue formed a living flag as they recited the pledge. In years to come, children across the nation—immigrant and native-born alike—would stand and recite that same pledge at the start of every school day.

To underscore the progress of the flag and the "inevitable triumph" of "white civilization" over Native Americans, the organizers invited several Sioux chiefs to the opening ceremonies. They made a brief appearance and then quietly left center stage, as a chorus sang "My Country 'Tis of Thee." A reporter for the *Chicago Tribune* noted, "Nothing in the day's occurrences appealed to the sympathetic patriotism so much as this fallen majesty slowly filing out of sight as the flags of all nations swept satin kisses through the air, waving congratulations to the cultured achievement and submissive admiration to a new world."[1]

That message also shaped the design of the exposition. The White City, as the fair was called, was supposed to represent the crowning achievement of American cultural and economic progress. In *The City of the Century*, historian Donald L. Miller writes:

> The spacious exhibition halls were arranged in sympathy with their natural surroundings and were conveniently interconnected by picturesque walkways and two and a half miles of watercourse. At almost every major point on the grounds, footsore sightseers could climb aboard a "swift and silent" electric launch or flag down a smaller battery-run boat—like hailing a cab—and head to the next spot on their guidebook agenda. The railroad that circled the grounds was the first in America to operate heavy, high-speed trains by electricity, and it ran on elevated tracks, posing no danger to pedestrians at a time when trains, trolleys, and cable cars killed more than four hundred people a year on the streets of Chicago.
>
> The streets and pavements of the White City were free of refuse and litter and patrolled by courteous Columbian Guards, drilled and uniformed like soldiers in the Prussian army; there was also a secret service force. . . . Every water fountain was equipped with a Pasteur filter, and the model sanitary system . . . worked flawlessly, converting sewage into solids and burning it, the ashes being used for road cover and fertilizer. There were no garish commercial signs, and with the concessionaires licensed and monitored, the fairgoers walked the grounds free from the nuisance of peddlers and confidence men, yet with the myriad pleasures of metropolitan life near at hand. The pavilions were vast department stores stocked with the newest consumer products, and in the course of a crowded day of sightseeing,

> visitors could stop at courteously staffed coffee shops, teahouses, restaurants, and beer gardens located at ground level or on rooftop terraces. The White City seemed to suggest a solution to almost every problem afflicting the modern city. . . .[2]

Problems that did not lend themselves to technological solutions were ignored. The week the exposition opened, a depression began in the United States. By 1894, over 16,000 businesses and 500 banks had failed. Hundreds of thousands of workers lost their jobs. The organizers paid no attention to these Americans other than to hire guards to keep them off the fairgrounds.

Officials also tried to eliminate dissent at the fair. Although many of the nation's leading thinkers, reformers, and religious leaders spoke at the exposition, audiences were not permitted to ask questions nor were the speakers allowed to address one another directly. Many Americans found the idea of a clean, sparkling city without controversy or poverty refreshing, even inspiring. The *Chicago Tribune* described the White City as "a little ideal world, a realization of Utopia . . . [foreshadowing] some far away time when the earth should be as pure, as beautiful, and as joyous as the White City itself." To Robert Herrick and other visitors to the Exposition, it was a magical place. He wrote: "The people who could dream this vision and make it real, those people . . . would press on to greater victories than this triumph of beauty—victories greater than the world had yet witnessed."[3]

At the nearby Midway Plaisance—a strip of land a mile long and 600 feet wide across from the White City, visitors encountered a lesson in "race science" and social Darwinism. Here they saw "living exhibits"—representatives of the world's "races" including Africans, Asians, and American Indians. The two German and two Irish villages were located nearest to the White City. Farther away and closer to the center of the Midway were villages representing the Middle East, West Asia, and East Asia. Then, wrote literary critic Denton J. Snider, "we descend to the savage races, the African of Dahomey and the North American Indian, each of which has its place" at the far end of the Plaisance. "Undoubtedly," he noted, "the best way of looking at these races is to behold them in the ascending scale, in the progressive movement; thus we can march forward with them starting with the lowest specimens of humanity, and reaching continually upward to the highest stage" so that "we move in harmony with the thought of evolution."

The fair's organizers promoted the idea that the "savage races" were dangerous by warning that "the [Dahomey] women are as fierce if not fiercer than the men and all of them have to be watched day and night for fear they may use their spears for other purposes than a barbaric embellishment of their dances." "The stern warning," writes anthropologist Lee Baker, "reinforced many Americans' fears that African Americans could not be trusted and were naturally predisposed to immoral and criminal behavior and thus kept away from white people through segregation."[4]

Some groups were outraged at the way they were presented at the fair. Emma Sickles, the chair of the Indian Committee of the Universal Peace Union, protested portrayals of Native Americans at the exhibition in *The New York Times* on October 8, 1893. Her letter states in part:

> Every effort has been put forth to make the Indian exhibit mislead the American people. It has been used to work up sentiment against the Indian by showing that he is either savage or can be educated only by government agencies. This would strengthen the power of everything that has been "working" against the Indians for years. Every means was used to keep the self-civilized Indians out of the Fair. The Indian agents and their backers knew well that if the civilized Indians got a representation in the Fair the public would wake up to the capabilities of the Indians for self-government and realize that all they needed was to be left alone.

African American leaders also protested. Frustrated and angry that "the Negroes' 'progress'" was ignored, two well-known African American activists, Frederick Douglass (Chapter 2) and Ida B. Wells, took matters in their own hands. They wrote and then distributed to fairgoers a pamphlet entitled *The Reason Why the Colored American Is Not in the World's Columbian Exposition*. As a concession to African Americans, organizers set aside a day in August as "Colored Jubilee Day." Although many blacks refused to participate, Douglass agreed to speak. He used the occasion to outline the progress made by African Americans since the Civil War despite injustices, acts of violence, and blatant persecution. He also lambasted fair organizers who fostered the belief "that our small participation in the World's Columbian Exposition is due either to our ignorance or to our want of public spirit."

CONNECTIONS

What is a fair? What is its purpose? How was the Columbian Exposition like other fairs you have attended? In what sense was it unique? What message did the exposition convey? What emotions did it prompt? Who was the intended audience? What do you think they learned from the fair?

How did the fair encourage patriotism? Build pride in the nation? Whom did the organizers see as part of that nation? Who seemed to lie beyond its universe of obligation? How did the organizers of the Columbian Exposition answer the question of Chapter 1: What do you do with a difference?

What does it mean to associate "whiteness" with being an American at a time of mass immigration?

Historian Donald I. Miller writes, "The White City seemed to suggest a solution to almost every problem afflicting the modern city." Identify some of those problems and the way each was solved in the White City. At whose expense were many of these problems solved?

Why do you think many Americans found the idea of city without controversy refreshing? What does this reading suggest about how those who disagreed with the majority could get heard in the late 1800s? How have new technologies affected our ability to voice our opinions? To have those opinions heard and respected?

To what extent was the "White City" a utopia? How is it like the "Masterpiece Society" described on pages 31 and 32? What differences seem most striking? Is either a democracy? A dystopia is the opposite of a utopia. To what extent was the Midway a dystopia? What lessons did it teach?

How did the fair's organizers define *civilization*? *Barbarism*? *Savagery*? What do those words mean to you? Record your definitions in your journal so that you can revise, expand, and deepen them as you continue to read.

Emma Sickles protested the way Indians were portrayed at the fair. Why was she outraged at the omission of "self-civilized" Indians? How does she seem to define the term *self-civilized*? What do you think it means to be "self-civilized"? If the so-called "inferior races" are able to "civilize" themselves, what questions do their efforts raise about social Darwinism? About the validity of the notion of "inferior" and "superior" races?

How did the organizers use "modern science," including Charles Darwin's theories (page 63), to reinforce old myths about "race"? How did they use "science" to not only rank the "races" but also justify those rankings?

Anthropologist Lee Baker believes that "the ethnological exhibits provided easy answers for Americans who were seeking ways to explain the violent chaos that erupted at the massacre at Wounded Knee in 1890, race and labor riots in 1892, terrorizing lynch mobs, and reports that African Americans composed the most criminal element in society." Research one event on Baker's list. What "easy answers" did the fair provide? Who might be attracted to those "easy answers"?

The Columbian Exposition was one of several "world's fairs" in the late 1800s and early 1900s. Research another fair, like the one in St. Louis in 1904 to mark the centennial of the Louisiana Purchase. What did fairgoers learn about human differences—about *us* and *them*?

---

1. Chicago *Tribune,* October 21, 1892.

2. Reprinted by permission of Simon & Schuster from *City of the Century: The Epic of the Making of America* by Donald L. Miller. Copyright © 1996 by Donald L. Miller.

3. *Memoirs of an American Citizen* by Robert Herrick. Harvard University Press, 1963, p. 147. Originally published in 1905.

4. *From Savage to Negro* by Lee D. Baker. University of California Press, 1998, p. 58.

"Progress," Civilization, and "Color-Line Murder"

Reading 4

In the late 1800s and early 1900s, the notion that the "white race" is "superior" to all others shaped the way many Americans viewed the world. Fairs like the World's Columbian Exposition simply confirmed what people already believed about *us* and *them.* Fears of miscegenation—the mixing of the "races"—were widespread and acts of violence against African Americans and other minorities were on the rise. Newspapers, magazines, and other publications too often viewed lynchings as "justice" served in the name of chivalry and the "protection of white women." It was a view supported by the nation's leading anthropologists and other scholars. Daniel G. Brinton, president of the International Congress of Anthropology and the American Association for the Advancement of Science in the 1890s, was among those who called for laws and educational reforms based on the "scientific fact" that African Americans were inferior to white Americans. In his most popular work, *Races and Peoples,* he argued:

> It cannot be too often repeated, too emphatically urged, that it is to the women alone of the highest race that we must look to preserve the purity of the type, and with it the claims of the race to be the highest. They have no holier duty, no more sacred mission, than that of transmitting in its integrity the heritage of ethnic endowment gained by the race throughout thousands of generations of struggle. . . . That philanthropy is false, that religion is rotten, which would sanction a white woman enduring the embrace of a colored man.[1]

Were lynchings an effort to protect "white women"? Journalist and social activist Ida B. Wells conducted an investigation to find out. In 1909, she reported her findings in a speech to members of the National Association for the Advancement of Colored People (NAACP):

> The lynching record for a quarter of a century merits the thoughtful study of the American people. It presents three salient facts: First, lynching is a color-line murder. Second, crimes against women are the excuse, not the cause. Third, it is a national crime and requires a national remedy.
>
> Proof that lynching follows the color line is to be found in the statistics which have been kept for the past twenty-five years. During the few years preceding this period and while frontier law existed, the executions showed a majority of white victims. Later, however, as law

courts and authorized judiciary extended into the far West, lynch law rapidly abated, and its white victims became few and far between. . . .

During the last ten years, from 1899 to 1908 inclusive, the number lynched was 959. Of this number, 102 were white, while the colored victims numbered 857. No other nation, civilized or savage, burns its criminals; only under the Stars and Stripes is the human holocaust possible. Twenty-eight human beings burned at the stake, one of them a woman and two of them children, is the awful indictment against American civilization—the gruesome tribute which the nation pays to the color line.

Why is mob murder permitted by a Christian nation? What is the cause of this awful slaughter? This question is answered almost daily: always that same shameless falsehood that "Negroes are lynched to protect womanhood." Standing before a Chautauqua assemblage, John Temple Graves, at once champion of lynching and apologist for lynchers, said, "The mob stands today as the most potential bulwark between the women of the South and such a carnival of crime as would infuriate the world and precipitate the annihilation of the Negro race." This is the never-varying answer of lynchers and their apologists. All know that it is untrue. The cowardly lyncher revels in the murder, then seeks to shield himself from public execration by claiming devotion to woman. But truth is mighty and the lynching record discloses the hypocrisy of the lyncher as well as his crime.

The Springfield, Illinois, mob rioted for two days, the militia of the entire state was called out, two men were lynched, hundreds of people driven from their homes, all because a white woman said a Negro assaulted her. A mad mob went to the jail, tried to lynch the victim of her charge, and, not able to find him, proceeded to pillage and burn the town and to lynch two innocent men. Later after the police had found that the woman's charge was false, she published a retraction, the indictment was dismissed, and the intended victim discharged. But the lynched victims were dead, hundreds were homeless, and Illinois was disgraced.

As a final and complete refutation of the charge that lynching is occasioned by crimes against women, a partial record of lynchings is cited; 285 persons were lynched for causes as follows: unknown cause, 92; no cause, 10; race prejudice, 49; miscegenation, 7; informing, 12; making threats, 11; keeping saloon, 3; practicing fraud, 5; practicing voodooism, 2; bad reputation, 8; unpopularity, 3; mistaken identity, 5; using improper language, 3; violation of contract, 1; writing insulting letter, 2; eloping, 2; poisoning horse, 1; poisoning well, 2; by white capes, 9; vigilantes, 14; Indians, 1; moonshining, 1; refusing evidence, 2; political causes, 5; disputing, 1; disobeying quarantine regulations, 2; slapping a child, 1; turning state's evidence, 3; protecting a Negro, 1; to prevent giving evidence, 1; knowledge of larceny, 1; writing letter to white woman, 1; asking white woman to marry, 1; jilting girl, 1; having small-pox, 1; concealing criminal, 2; threatening political exposure, 1; self-defense, 6; cruelty, 1; insulting language to woman, 5; quarreling with white man, 2; colonizing Negroes, 1; throwing stones, 1; quarreling, 1; gambling, 1.

Is there a remedy, or will the nation confess that it cannot protect its protectors at home as well as abroad? Various remedies have been suggested to abolish the lynching infamy; but year after year, the butchery of men, women, and children continues in spite of plea and protest. Education is suggested as a preventative, but it is as grave a crime to murder an ignorant man, as it is a scholar. True, few educated men have been lynched, but the hue and cry once started stops at no bounds, as was clearly shown by the lynchings in Atlanta, and in Springfield, Illinois.

Agitation, though helpful, will not alone stop the crime. Year after year statistics are published, meetings are held, resolutions are adopted. And yet lynchings go on. . . . The only certain remedy is an appeal to law. Lawbreakers must be made to know that human life is sacred and that every citizen of this country is first a citizen of the United States and secondly a citizen of the state in which he belongs. This nation must assert itself and protect its federal citizenship at home as well as abroad. The strong men of the government must reach across state lines whenever unbridled lawlessness defies state laws, and must give to the individual under the Stars and Stripes the same measure of protection it gives to him when he travels in foreign lands. Federal protection of American citizenship is the remedy for lynching. . . .

In a multitude of counsel there is wisdom. Upon the grave question presented by the slaughter of innocent men, women, and children there should be an honest, courageous conference of patriotic, law-abiding citizens anxious to punish crime promptly, impartially, and by due process of law, also to make life, liberty, and property secure against mob rule.

Time was when lynching appeared to be sectional, but now it is national—a blight upon our nation, mocking our laws and disgracing our Christianity. "With malice toward none but with charity for all," let us undertake the work of making the "law of the land" effective and supreme upon every foot of American soil—a shield to the innocent; and to the guilty, punishment swift and sure.[2]

CONNECTIONS

How does Ida B. Wells define *lynching*? In her view, what is the relationship between lynching and the way a community defines its universe of obligation?

In the 1920s, composer Arnold Schoenberg witnessed antisemitism, a form of racism, in Germany. He asked, "But where is antisemitism to lead to if not to acts of violence?" How would you answer his question? How are racism and violence linked to the way a nation defines its universe of obligation? What does Wells suggest about the role of "the mob" in the way the two are linked?

How does Wells define such words as *civilization*, *barbarism*, *citizenship*, and *liberty*? How do her definitions differ from those of the organizers of the Columbian Exposition? What relationship does she see between individual liberties and the law? Compare and contrast her views with those expressed by Mark Sullivan in Reading 2. How do you account for differences?

How does Wells use statistics to educate the public and sway public opinion? What do the numbers reveal that words could not convey? What other techniques does she use to make her case? Which is most effective?

Like individuals, communities and even nations have identities. Use the information in this reading to create an identity chart for the United States at the turn of the 20th century. Include the words or phrases the nation uses to describe itself as well as the ones that others attach to it. How did the nation seem to define its universe of obligation? Who was outside that universe? As you continue reading this chapter, add to the nation's identity chart.

For more information on the life and work of Ida B. Wells, see *Choosing to Participate: A Critical Examination of Citizenship in American History*. Also available from the Facing History Resource Center is a documentary entitled *A Passion for Justice: The Life of Ida B. Wells*.

---

1. *Races and Peoples* by Daniel Brinton. Hodges, 1890, p. 287.
2. Reprinted in *In Our Own Words* edited by Robert Torricelli and Andrew Carroll. Pocket Books, 1999, pp. 22-25.

Doors to Opportunity

Reading 5

Progressives believed in the power of education to "civilize," "uplift," and "Americanize." In every state in the nation, they lobbied for laws that required children to attend school until at least the age of 14. Partly as a result of their efforts, school enrollment in the United States increased by more than 600 percent, from about 200,000 students in 1880 to over 1.5 million by 1920.

Yet even as progressive reformers worked to expand educational opportunities, many were uncertain that all children could benefit from schooling. Increasingly some argued that education should be made available only to those with a large "endowment" of certain qualities like "character, disposition, energy, intellect, or physical power"—qualities that "go towards the making of civic worth." They insisted that placing groups (based on "race," class, or gender) in the right educational "track" or even in a special school to train them for their "rightful place" in society was the most efficient use of taxpayers' money.

The experiences of two young immigrants reveals how "race" shaped the kind of education available to many Americans in the late 1800s. Mary Antin, a young Jewish immigrant from Poland, came to Boston with her mother and siblings in 1894. They reunited with Antin's father who had arrived earlier to find a job and establish a home. She writes in her autobiography:

> Education was free. That subject my father had written about repeatedly, as comprising his chief hope for us children, the essence of American opportunity, the treasure that no thief could touch, not even misfortune or poverty. It was the one thing that he was able to promise us when he sent for us, surer, safer, than bread or shelter.
>
> On our second day [in America] I was thrilled with the realization of what this freedom of education meant. A little girl from across the alley came and offered to conduct us to school. My father was out, but we five [children] between us had a few words of English by this time. We knew the word *school*. We understood. This child, who had never seen us till yesterday, who could not pronounce our names, who was not much better dressed than we, was able to offer us the freedom of the schools of Boston! No application made, no questions asked, no examinations, rulings, exclusions; no machinations, no fees. The doors stood open for every one of us. The smallest child could show us the way.
>
> This incident impressed me more than anything I had heard in

> advance of the freedom of education in America. It was a concrete proof—almost the thing itself. One had to experience it to understand it.[1]

Ten years before the Antins came to the United States, another immigrant family tried to enroll their daughter at a public school in San Francisco only to be turned away. Principal Jennie Hurley explained that the Board of Education did not allow children of Chinese descent to attend the city's public schools. In 1884, the Tapes sued the principal in a case known as *Tape v. Hurley* for denying Mamie, their 8-year-old daughter, an education. Hurley and other school officials defended the child's exclusion by pointing to a clause in the California constitution describing the Chinese as "dangerous to the well-being of the state." Therefore, they argued, the city was not obligated to educate the Chinese.

Despite the school board's argument, the courts ruled in the Tapes' favor, citing a state law requiring that "all children" be admitted to school; only "children of filthy or vicious habits," or "children suffering from contagious or infectious diseases" could be excluded. Mamie Tape had the "same right to enter a public school" as any other child. Officials responded to the ruling by establishing a special public school just for Mamie Tape and any other Chinese child who wished to attend. Outraged, Mary Tape, young Mamie's mother, wrote a letter in newly learned English to the Board of Education in April of 1885:

> DEAR SIRS: I see that you are going to make all sorts of excuses to keep my child out of the Public Schools. Dear sirs, Will you please to tell me! Is it a disgrace to be Born a Chinese? Didn't God make us all!!!! What right have you to bar my children out of the school because she is a Chinese Decend. There is no other worldly reason that you could keep her out, except that. I suppose, you all goes to church on Sundays! Do you call that a Christian act to compel my little children to go so far to a school that is made on purpose for them. My children don't dress like the other Chinese. They look just as phunny amongst them as the Chinese dress in Chinese look amongst you Caucasians. Besides, if I had any wish to send them to a Chinese school I could have sent them two years ago without going to all this trouble. You have expended a lot of Public money foolishly, all because of one poor little Child. Her playmates is all Caucasians ever since she could toddle around. If she is good enough to play with Them! Then is she not good enough to be in the same room and study with them? You had better come and see for yourselves. See if

> the Tapes is not the same as other Caucasians except in features. It seems not matter a Chinese may live and dress so long as you know they Chinese. Then they are hated as one. There is not any right or justice for them.
>
> You have seen my husband and child. You told him it wasn't Mamie Tape you object to. If it were not Mamie Tape you object to, then why didn't you let her attend the school nearest her home! Instead of first making one pretense of some kind to keep her out? It seems to me Mr. Moulder has a grudge against this Eight-year-old Mamie Tape. I know they is no other child I mean Chinese child! Care to go to your public Chinese school. May you Mr. Moulder, never be persecuted like the way you have persecuted little Mamie Tape. Mamie Tape will never attend any of the Chinese schools of your making! Never!!! I will let the world see sir What justice there is When it is govern by the Race prejudice men! Just because she is of the Chinese decend, not because she don't dress like you because she does. Just because she is decended of Chinese parents I guess she is more of a American than a good many of you that is going to prevent her being Educated.[2]

School officials ignored Mary Tape's appeal. So did the California courts. At the turn of the 20th century, both state and federal courts supported the idea of "separate but equal" schools for children of "inferior races." Mamie Tape had to attend a segregated school or not be educated at all.

In 1906, San Francisco school officials decided to send students of Japanese and Korean descent to Mamie Tape's school. Over 1,200 Japanese parents in the city responded to the announcement with a lawsuit attacking segregation. The Japanese consul in California wrote a formal letter of protest to government officials in Washington, D.C. President Theodore Roosevelt, eager to maintain good relations with Japan, sent a member of his cabinet to San Francisco to encourage the school board to withdraw its order. Although the Japanese were pleased with Roosevelt's stand, San Francisco school officials were unimpressed. In the end, however, they agreed to a political compromise. The city would allow Japanese children to attend all-white schools if the federal government sharply limited the number of Japanese laborers who could enter the United States each year. In the years that followed, those limits were tightened further so that fewer and fewer Japanese immigrants could settle in the nation.

CONNECTIONS

What do Mary Antin's recollections add to our understanding of what it means to be within a nation's universe of obligation? What does Mary Tape's outrage suggest about what it means to lie beyond that universe of obligation?

In the late 1800s, many white Americans claimed that the Chinese could not be assimilated—that is, acquire American values and traditions. They were "too different." How does Mary Tape shatter those stereotypes? The right to petition government officials is central to democracy. It is guaranteed in the first amendment to the Constitution. How does Tape show her understanding of that principle? What other democratic principles are reflected in her writing?

How do you explain the failure of public officials to respond to Tape's letter? If they had responded, how might they have answered the questions she raises? How would you answer them?

China in the late 1800s and early 1900s was a weak nation torn by war. During those years, Japan was becoming a modern industrialized nation with a strong military. How did this reality shape the way the Chinese and the Japanese were treated in the United States? To what extent did their treatment reflect myths about race and racial differences?

---

1. *The Promised Land* by Mary Antin. Houghton Mifflin, 1910, pp. 185-186.
2. Quoted in *Letters of a Nation* edited by Andrew Carroll. Kodansha America, Inc., 1997, pp. 185-186.

Taking Up the "White Man's Burden"

Reading 6

Questions about "race" and membership shaped not only American life but also the nation's foreign policy. In 1898, just after the Spanish American War, Americans discussed the future of the territories it acquired as a result of its victory in that war.

The United States declared war on Spain on April 25, 1898, after the U.S. battleship *Maine* blew up in Havana harbor in Cuba. Although investigators never determined the cause of the explosion, American newspapers were quick to call the disaster "wholesale murder" and blame Spain, which was trying to put down a rebellion in Cuba, then a Spanish colony. Many Americans supported the war because they believed it would result in independence for Cuba. Others regarded the war as an opportunity to gain territory abroad. As early as February 1898, Theodore Roosevelt, then assistant secretary of the Navy, sent secret orders to Commodore George Dewey, head of the American fleet in the Pacific. Roosevelt ordered Dewey to take the Philippines, which then belonged to Spain, as soon as war began. Dewey obeyed.

By August, the war was over and the United States controlled not only Cuba but also the Philippines, Guam, and Puerto Rico. The United States had to decide what to do with these islands. The debate focused on the Philippines. There was general agreement that Cuba, which had been fighting for years to overthrow Spanish rule, ought to be independent, although the American government placed limits on the new nation's freedom. There was also little doubt about the future of Guam and Puerto Rico, even though Puerto Ricans had been virtually independent of Spain for a number of years. Both now were under American rule. Many Americans were also eager to keep the Philippines. Others noted that the Filipinos had been fighting for their independence long before the war began. It did not seem right to give Cuba its freedom and make the Philippines a colony.

Each side in the debate used American principles to support its point of view. Each also relied on racist thinking. Those who favored intervention argued that the nation had a responsibility to not only rule "inferior races" but also "educate," "uplift," and "civilize" them. Among these Americans was Senator Albert J. Beveridge who argued:

> Think of the thousands of Americans who will pour into Hawaii and Puerto Rico when the republic's laws cover those islands with

justice and safety! Think of the tens of thousands of Americans who will invade mine and field and forest in the Philippines when a liberal government, protected and controlled by this republic, if not the government of the republic itself, shall establish order and equity there! Think of the hundreds of thousands of Americans who will build a soap-and-water, common-school civilization of energy and industry in Cuba, when a government of law replaces the double reign of anarchy and tyranny! . . .

What does all this mean for every one of us? It means opportunity for all the glorious young manhood of the republic—the most virile, ambitious, impatient, militant manhood the world has ever seen.[1]

Anti-imperialists like Mark Twain strongly disagreed. He based his argument on the Declaration of Independence: "The hearts of men are about alike, all over the world, no matter what their skin-complexions may be."[2] Senator Ben Tillman of South Carolina, also an anti-imperialist, based his opposition to expansion on "race science." He argued that white southerners "understand and realize what is to have two races side by side that can not mix or mingle without deterioration and injury to both and the ultimate destruction of the civilization of the higher." A British writer named Rudyard Kipling participated in the debate by addressing Americans in a poem that was widely quoted at the time:

Take up the White Man's burden—
  Send forth the best ye breed—
Go bind your sons to exile
  To serve your captives' need;
To wait in heavy harness,
  On fluttered folk and wild—
Your new-caught, sullen peoples,
  Half-devil and half-child.

Kipling ended his poem with the following verse:

Take up the White Man's burden—
  Have done with childish days—
The lightly proffered laurel,
  The easy, ungrudged praise.
Comes now, to search your manhood
  Through all the thankless years,
Cold, edged with dear-bought wisdom,
  The judgment of your peers![3]

The ideas expressed in the poem troubled many African Americans, particularly those who had fought in the Spanish American War to show their loyalty, courage, and idealism at a time when others portrayed them as inferior, cowardly, and immoral. Now some were uncomfortable with the consequences of their sacrifices. In response to Kipling's poem, H. T. Johnson, a black clergyman and editor of the *Christian Recorder*, wrote:

> Pile on the Black Man's Burden.
> 'Tis nearest at your door;
> Why need long bleeding Cuba,
> or dark Hawaii's shore?
> Hail ye your fearless armies,
> Which menace feeble folks
> Who fight with clubs and arrows
> And brook your rifle's smoke.[4]

In 1899, the Senate approved by a single vote a treaty that placed the Philippines under American rule. Filipinos responded with a revolt that took 84,000 American soldiers over four years to end. In 1904, the United States marked its victory over the Filipinos at a world's fair held in St. Louis, Missouri, to celebrate the 100th anniversary of the Louisiana Purchase—the nation's first major expansion beyond the Mississippi River. Officials saw the fair as an opportunity to educate Americans about the nation's objectives in the Philippines by creating a special exhibition that contrasted "educated" Filipinos with "backward tribes" in need of "civilization." The aim was to show that American imperialism was not just an effort to gain land and wealth but also to educate, "uplift," civilize, and Christianize a "primitive people."

To show the positive effects of American civilization on native peoples, officials imported several hundred members of the Philippine Scouts and the Constabulary. At the fair, the two groups performed drills and other military maneuvers to show that "savages" could be "transformed" through education. Also included at the fair were the Igorots, "an uncivilized tribe" that revealed how much work still needed to be done. Attracted by stories of "naked savages" who dined on "dog meat," thousands of visitors flocked to see the Igorots and even have their pictures taken with them. They were such an attraction that they were showcased at other fairs in the years that followed, despite the protests of many Filipinos. A Filipino newspaper explained:

> As Americans may have no better sources of information, they believe that the majority of the Filipinos are like the [Igorots]. There are many of our students and countrymen who have been asked the

following questions from badly informed Americans: "Since when have you used coats?" "Do your shoes hurt your feet?" and whether there are many Filipinos who wear clothes, etc., etc. And as the United States government maintains that its mission is one of education, the belief grows that we Filipinos are savages whom the nephews of Uncle Sam are here to civilize.

When the exposition was held at St. Louis, we energetically opposed the exhibition of non-Christian tribes; the effect on the opinion in the United States verified our fears. Again we opposed the sending of them to [the fair in] Portland. We were equally unsuccessful in this.

It does little good to send honorary commissioners, delegates, students, etc., to America; the general opinion continues that they are exceptional samples and that the masses are still "savages." Congressional delegations and travelers . . . may come; but what are

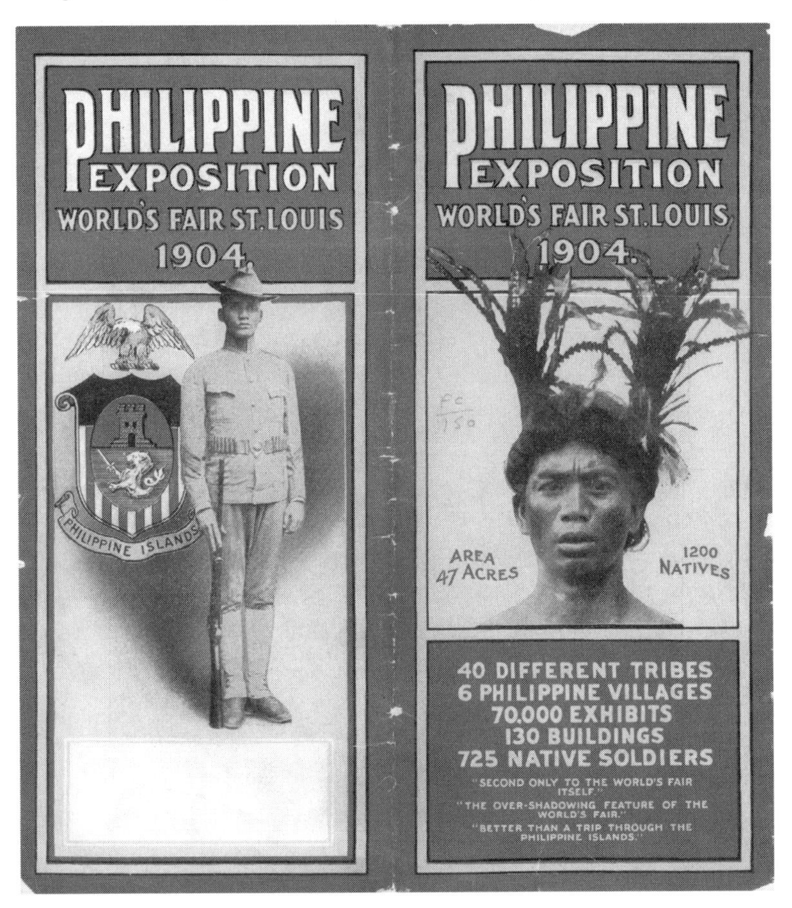

The cover of a booklet promoting the Philippine Exposition at the St. Louis World's Fair. Notice the contrast between the Philippine scout and the Igorot.

> these drops in the midst of that ocean of American impression formed by the sight of these non-Christian tribes? Besides this, those who come here and return to America are not all sincerely actuated by wishes for the highest good of the Filipinos. How then can the truth be established which political interests are interested to conceal?[5]

CONNECTIONS

How do the individuals quoted in this reading seem to define such terms as *civilization*, *liberty*, *democracy*, and *barbarism*? Compare and contrast their views with those expressed in earlier readings.

How do the individuals quoted in the reading seem to define the nation's universe of obligation? Who is part of that universe? Who lies beyond it? What other ideas influence opinions on whether the United States should acquire colonies?

How important were exhibitions like the St. Louis World's Fair of 1904 in shaping public opinion about the issues of the day? What is the difference between reading about Filipinos and how they live and seeing an "authentic re-creation" of their way of life? What does the protest in the Filipino newspaper suggest about the challenges of altering public opinion?

Emma Sickles objected to the fact that "self-civilized Indians" were excluded from the World's Columbian Exposition in Chicago (Reading 3). How might her comments apply to the portrayal of Filipinos at the St. Louis fair? A number of websites contain photographs, magazine articles, and other artifacts from the St. Louis fair. Use them to compare and contrast the treatment of American Indians and other groups at the fair with that of the Filipinos. What similarities do you notice? How do you account for differences?

In an address to African Americans in 1900, Thomas Wentworth Higginson, William Lloyd Garrison, and George S. Boutwell—three white veterans of the abolitionist movement and the Civil War—expressed their concerns about the consequences of American involvement in the Philippines:

> Every day in the Philippines is already training our young American soldiers to the habit of thinking that the white man, as such, is the rightful ruler of all other men. This is seen, for instance, in the fact that these very soldiers, in writing home letters from the seat of

war, describe the inhabitants of the Philippines, more and more constantly, as "n—s"; thus giving a new lease of life to a word which was previously dying out among us.

. . . In other words, freedom is to become . . . a matter of complexion. If this doctrine is to prevail, what hope is there for the colored race in the United States? The answer is easy; there is in that case no hope at all.[6]

Why do the three believe that both black and white Americans ought to dread "the habit of thinking that the white man . . . is the rightful ruler of all other men"? A growing belief that freedom is "a matter of complexion"? What is the danger to each group? To democracy?

---

1. Quoted in *The American Spirit* edited by Thomas A. Bailey. D.C. Heath, 1963, pp. 609-610.
2. Quoted in *Barbarian Virtues* by Matthew Frye Jacobson. Hill and Wang, 2000, p. 230.
3. From "The White Man's Burden" by Rudyard Kipling. *McClurg's Magazine* 12, February 1899.
4. Quoted in *Barbarian Virtues* by Matthew Frye Jacobson. Hill and Wang, 2000, p. 257.
5. "How the Filipinos Feel about the Exhibition of the Igorots in the United States." *The Public*, vol. 8, March 3, 1906.
6. Higginson, Thomas Wentworth, William Lloyd Garrison, and George S. Boutwell. "Address to the Colored People of the United States." *Voice of Missions* 8 (Nov. 1, 1900). *http://www.boondocksnet.com/ailtexts/adcol926.html* In Jim Zwick, ed., *Anti-Imperialism in the United States, 1898-1935. http://www.boondocksnet.com/ail98-35.html*

Reading 7

Race was not the only issue that divided Americans at the turn of the 20th century. Americans were also divided by social class. In 1890, just one percent of American families owned 51 percent of the nation's real estate and personal property. The poorest 44 percent owned a little over 1 percent.[1] That disparity troubled many people. As early as 1879, sociologist Henry George wrote that despite the nation's "prodigious increase in wealth-producing power . . . it becomes no easier for the masses of our people to make a living. On the contrary, it is becoming harder. The gulf between the employed and the employer is becoming wider; social contrasts are becoming sharper; as liveried carriages appear, so do barefooted children."[2]

The pictures that accompany this reading offer a visual perspective on the disparities between the lives of the rich and the poor. That gap is also evident in the childhood memories of Americans who grew up in the late 1800s and early 1900s. Novelist Edith Wharton came from a socially prominent New York City family. In the 1930s, she recalled the foods of her childhood:

> My father had inherited from his family a serious tradition of good cooking. . . . My mother, if left to herself, would probably not have been much interested in the pleasures of the table. My father's Dutch blood accounted for his gastronomic enthusiasm; his mother, who was a Schermerhorn, was reputed to have been the best cook in New York. But to know about good cooking was a part of every young wife's equipment, and my mother's favorite cookery books (Francatelli's and Mrs. Leslie's) are thickly interleaved with sheets of yellowing note paper, on which, in a script of ethereal elegance, she records the making of "Mrs. Joshua Jones's scalloped

A detail of the painting "Daughters of Edward Darley Boit" by John Singer Sargent.

> oysters with cream," "Aunt Fanny Gallatin's fried chicken," "William Edgar's punch," and the special recipes of our two famous Negro cooks, Mary Johnson and Susan Minnerman. . . . Mary Johnson, a gaunt towering woman of a rich bronzy black, with huge gold hoops in her ears, and crisp African crinkles under vividly patterned kerchiefs; Susan Minnerman, a small smiling mulatto, more quietly attired, but as great a cook as her predecessor.
>
> Ah, what artists they were! How simple yet sure were their methods—the mere perfection of broiling, roasting and basting—what an unexampled wealth of material, vegetable and animal, their genius had to draw upon! Who will ever again taste anything in the whole range of gastronomy to equal their corned beef, their boiled turkeys with stewed celery and oyster sauce, their fried chickens, broiled red-heads, corn fritters, stewed tomatoes, rice griddle cakes, strawberry short-cake and vanilla ices? I am now enumerating only our daily fare, that from which even my tender years did not exclude me; but when my parents "gave a dinner," and terrapin and canvas-back ducks, or (in their season) broiled Spanish mackerel, soft-shelled crabs with a mayonnaise of celery, and peach-fed Virginia hams cooked in champagne (I am no doubt confusing all the seasons in this allegoric evocation of their riches), lima-beans in a cream, corn soufflés, and salads of oyster-crabs, poured in varied succulence from Mary Johnson's lifted cornucopia—ah, then, the gourmet of that long-lost day, when cream was cream and butter butter and coffee coffee, and meat fresh every day, and game hung just for the proper number of hours, might lean back in his chair and murmur "Fate cannot harm me" over his cup of Moka [a coffee made from costly and aromatic beans] and his glass of authentic Chartreuse [a liqueur].
>
> I have lingered over these details because they formed a part—a most important and honorable part—of that ancient curriculum of house-keeping which, at least in Anglo-Saxon countries, was so soon to be swept aside by the "monstrous regiment" of the emancipated: young women taught by their elders to despise the kitchen and the linen room, and to substitute the acquiring of University degrees for the more complex art of civilized living. The movement began when I was young, and now that I am old, and have watched it and noted its results, I mourn more than ever the extinction of the household arts.[3]

Although Wharton's family was not fabulously wealthy, few Americans could afford to set as elaborate a dinner table as her parents did. According to the 1900 census, two-thirds of all male workers over the age of 16 earned less than $12.50 a week. And about one of every four of those workers could expect to be laid off at a moment's notice. Although a dollar in 1900 bought far more than it does today, surviving on less than two dollars a day was a struggle. Like Wharton, journalist Zalmen Yoffeh made his home in New York City. His parents were immigrants from Eastern Europe who struggled to make a living. Yoffeh recalls how his mother practiced the "household arts."

> With . . . one dollar a day [our mother] fed and clothed an ever-growing family. She took in boarders. Sometimes this helped; at other times it added to the burden of living. Boarders were often out of work and penniless; how could one turn a hungry man out? She made all our clothes. She walked blocks to reach a place where meat was a penny cheaper, where bread was a half-cent less. She collected boxes and old wood to burn in the stove instead of costly coal. Her hands became hardened and the lines so begrimed that for years she never had perfectly clean hands. One by one she lost her teeth—there was no money for dentists—and her cheeks caved in. Yet we children always had clean and whole clothing. There was always bread and butter in the house, and, wonder of wonders, there was usually a penny apiece for us to buy candy with. On a dollar and a quarter we would have lived in luxury.[4]

Detail from "Street children, New York City, c. 1908" by Lewis Hine.

Sammy Aaronson, a prizefighter, came from an even poorer immigrant family. He later recalled:

> Eating was always a struggle. We ate when we had food in the house and our diet would give a social service worker the horrors. Meat soup was a big thing and we sometimes could have it once a week. Outside of that, the only hot food we ever had was potatoes. I never tasted anything like steak or roast beef or lamb chops until I was sixteen years old. We lived on pumpernickel [bread], herring,

bologna ends, and potatoes. The whole family could eat for fifteen or twenty cents a day, sometimes less. Mom would send me over to the delicatessen on Hester Street where we could get pumpernickel the size of a steering wheel for a dime. We paid a penny a herring and two took care of the whole family. Another penny bought three pounds of potatoes. We always had the meat soup on Friday nights. It was made up of leftovers and ends and bones which the butcher sold for six cents a pound instead of throwing away. Three pounds was plenty for a meal for us.[5]

CONNECTIONS

Every picture tells a story. Look carefully at each of the pictures included in this reading. What story does each tell? Who is telling the story? What does each picture add to your understanding of the gap between rich and poor? Of why people like Henry George feared the possible consequences of that disparity?

The word *disparity* comes from a Latin word that means separate or distinct. List some of the disparities described in this reading. What questions do they raise? Why might these questions trouble many Americans?

Florence Harriman, whose family made its fortune in railroads, described "high society" as "pink frosting on a cake—a cake in a world that hungered for bread. . . . But that is only a mood. On the whole I have loved balls, garden parties, and hunting, as a pony loves his paddock. I cannot be solemn about the snobbery and the wastefulness. . . . The truth is that snobbery is not so wicked,—it is usually very, very dull, and as for wastefulness, if one believes in private property at all, I think that the . . . Balls that added to the gaiety of nations and set money in circulation were far more pious enterprises than unostentatious hoarding."[6] What does Harriman suggest that it means to be among the few to enjoy "cake with pink frosting" in a world that "hungers for bread"? Why do you think she dismisses her concerns as a "mood"? How does she seem to defend the balls and parties that define her world? How is the relationship she describes between rich and poor similar to the one implied in Edith Wharton's account?

According to Wharton, what is "civilized living"? Why does she seem to associate it with "Anglo-Saxon countries"? What value does she place on "civilized living"? What is the "monstrous regiment" that threatens it? Does that threat come from the poor or from others in her social class?

In the late 1800s, about 20 percent of all children between the ages of 10 and 14 held jobs. By the age of 14, half of the nation's children worked in factories, mines, and other businesses. How do accounts like those of Yoffeh and Aaronson help us understand why few poor children were able to attend school at that time? In 1900, the U.S. Navy turned away many young volunteers from poor families because they were physically unfit. How do accounts like those of Yoffeh and Aaronson help us understand why some men raised in poverty lacked the strength necessary for military service?

---

1. Cited in *A Very Different Age: Americans of the Progressive Era* by Steven J. Diner. Hill & Wang, 1998, p. 4.
2. *Poverty and Progress* by Henry George. Quoted in *Pivotal Decades: The United States, 1900-1920* by John Milton Cooper, Jr. W.W. Norton & Co., 1990, p. 10.
3. From Chapter 3 of *A Backward Glance* by Edith Wharton. Reprinted in *The Faber Book of America* edited by Christopher Ricks and William L. Vance. Faber and Faber, 1992, pp. 315-316.
4. "The Passing of the East Side" in *Menorah Journal*, December, 1929. Reprinted in *How We Live* edited by Irving Howe and Kenneth Libo. Richard Marek Publishers, 1979, p. 43.
5. *High as My Heart* by Sammy Aaronson and Albert S. Hirshberg. Coward, McCann & Geogheagan, Inc., 1957. Reprinted in *How We Live* edited by Irving Howe and Kenneth Libo. Richard Marek Publishers, 1979, p. 44.
6. Quoted in *The Rise of Industrial America* by Page Smith. McGraw Hill, 1984, pp. 859-860.

Reading 8

Many progressive reformers were deeply troubled by the widening gap between the rich and the poor. Fearful that huge disparities not only in wealth but also in opportunity might lead to revolution, they proposed a variety of laws and other reforms. Much of their effort focused on the nation's largest cities, where the gap between rich and poor was the most visible.

Jacob Riis, a Danish immigrant who worked as a police reporter in New York City, expressed the views of many of these progressives when he wrote, "Government by the people must ever rest upon the people's ability to govern themselves, upon their intelligence and public spirit. The slum stands for ignorance, want, unfitness, for mob-rule in the day of wrath. This at one end. At the other, hard-heartedness, indifference, self-seeking, greed. It is human nature. We are brothers whether we own it or not, and when the brotherhood is denied in Mulberry Street [one of New York's poorest neighborhoods] we shall look vainly for the virtue of good citizenship on Fifth Avenue [one of the city's richest neighborhoods]."[1]

In 1890, Riis published a detailed study of the tenements of New York City to explain why the slums on "Mulberry Street" ought to matter to those who lived on "Fifth Avenue."

> Long ago it was said that "one half of the world does not know how the other half lives." That was true then. It did not know because it did not care. The half that was on top cared little for the struggles, and less for the fate of those who were underneath, so long as it was able to hold there and keep its own seat. There came a time when the discomfort and crowding below were so great, and the consequent upheavals so violent, that it was no longer an easy thing to do, and then the upper half fell to inquiring what was the matter. Information on the subject has been accumulating rapidly since, and the world has had its hands full answering for its old ignorance.[2]

Riis went to describe early efforts to identify the "nursery of crime." After noting "younger criminals seem to come almost exclusively from the worst tenement districts," he concluded that the "boundary line of the Other Half lies through the tenements."

> The boundary line lies there because, while the forces for good

on one side vastly outweigh the bad—if it were not well otherwise—in the tenements all the influences make for evil; because they are the hotbed of the epidemics that carry death to rich and poor alike; the nurseries of pauperism and crime that fill our jails and police courts; that throw off a scum of forty thousand human wrecks to the island asylums and workhouses year by year; beggars to prey upon our charities; that maintain a standing army of ten thousand tramps with that implies; because, above all, they touch the family life with deadly moral contagion. This is their worst crime, inseparable from the system. That we have to own it as the child of our own wrong does not excuse it, even though it gives its claim upon our utmost patience and tenderest charity.

What are you going to do about it? is the question of today.[3]

Riis's book contains detailed descriptions of New York City's worst tenements, photographs of the individuals who lived there, and statistics drawn from such sources as the U.S. Census, police reports, and the city's health department. He was not the only progressive reformer to rely on reports, studies, and tables to advocate for change. In 1910, Jane Addams described her efforts to improve sanitation in the area around Hull House, the settlement that she founded with Ellen Gates Starr in Chicago after visiting the world's first settlement house in London. Like its British counterpart, Hull House was a place where newcomers to the city—both immigrants and the native-born—could find advice, childcare, English classes, lectures, clubs, and political groups. It was also a place where newcomers learned to participate positively in the life of their community.

During our first three years on Halsted Street, we had established a small incinerator at Hull House and we had many times reported the untoward conditions of the ward to the city hall. We had also arranged many talks for the immigrants, pointing out that although a woman may sweep her own doorway in her native village and allow the refuse to innocently decay in open air and sunshine, in a crowded city quarter, if the garbage is not properly collected and destroyed, a tenement-house mother may see her children sicken and die, and that the immigrants must therefore not only keep their own houses clean, but must also help the authorities to keep the city clean.

Possibly our efforts slightly modified the worst conditions, but they still remained intolerable, and the fourth summer . . . we began a systematic investigation of the city system of garbage collection, both as to its efficiency in other wards and its possible connection with the death rate in the various wards of the city.

. . . . The members [of the Hull House Women's Club] came together . . . in quite a new way that summer when we discussed with them the high death rate so persistent in our ward. After several meetings devoted to the subject, despite the fact that the death rate rose highest in the congested foreign colonies and not in the streets in which most of the Irish American club women lived, twelve of their number undertook in connection with the residents, to carefully investigate the condition of the alleys. During August and September the substantiated reports of violations of the law sent in from Hull House to the health department were one thousand and thirty-seven. For the club woman who had finished a long day's work of washing or ironing followed by the cooking of a hot supper, it would have been much easier to sit on her doorstep during a summer evening than to go up and down ill-kept alleys and get into trouble with her neighbors over the condition of their garbage boxes. It required both civic enterprise and moral conviction to be willing to do this three evenings a week during the hottest and most uncomfortable months of the year. Nevertheless, a certain number of women persisted, as did the residents, and three city inspectors in succession were transferred from the war because of unsatisfactory services. Still the death rate remained high and the condition seemed little improved throughout the next winter. In sheer desperation, the following spring when the city contracts were awarded for the removal of garbage, with the backing of two well-known businessmen, I put in a bid for the garbage removal of the nineteenth ward. My paper was thrown out on a technicality but the incident induced the mayor to appoint me the garbage inspector of the ward.

The salary was a thousand dollars a year, and the loss of that political "plum" made a great stir among the politicians. The position was no sinecure whether regarded from the point of view of getting up at six in the morning to see that the men were early at work; or of following the loaded wagons, uneasily dropping their contents at intervals, to their dreary destination at the dump; or of insisting that the contractor must increase the number of his wagons from nine to thirteen and from thirteen to seventeen, although he assured me that he lost money on every one and that the former inspector had let him off with seven; or of taking careless landlords into court because they would not provide the proper garbage receptacles; or of arresting the tenant who tried to make the garbage wagons carry away the contents of his stable.

With the two or three residents who nobly stood by, we set up six of those doleful incinerators which are supposed to burn garbage with the fuel collected in the alley itself. The one factory in town which could utilize old tin cans was a window weight factory, and we deluged that with ten times as many tin cans as it could use—much less would pay for. We made desperate attempts to have the dead animals removed by the contractor who was paid most liberally by the city for that purpose but who, we slowly discovered, always made the police ambulances do the work, delivering the carcasses upon freight cars for shipment to a soap factory in Indiana where they were sold for a good price although the contractor himself was the largest stockholder in the concern.[4]

CONNECTIONS

What is the message of Jacob Riis's writing? How does he use adjectives and descriptive nouns to underscore that message? What is the effect of such words and phrases as "hotbed of the epidemics," "nurseries of pauperism and crime," and "deadly moral contagion"?

What is the message of Jane Addams's account? How does she use adjectives and descriptive nouns to underscore that message? What is the effect of such words and phrases as "systematic investigation," "substantiated reports of violations of the law," and "civic enterprise and moral conviction"? How do the stories she tells underscore her message? Who is her intended audience?

"Woman's place is in the Home," wrote suffragist Rheta Childe Dorr in 1910, "but Home is not contained within the four walls of an individual home. Home is the community."[5] How do her remarks help us understand why Jane Addams and her staff devoted so much time and effort to garbage collection? What do these efforts suggest about how they defined their universe of obligation?

Although Riis offers no specific solution to the problem he describes, he hints at remedies. What are those remedies? How do they differ from the ones Jane Addams seems to advocate? Find out how sociologists, journalists, and others view similar problems today. To what extent have attitudes toward the poor changed? To what extent are they unchanged? Which remedies might be attractive to a "social Darwinist" like William Graham Sumner (Chapter 3)? A eugenicist like Charles Davenport (Chapter 3)? Which remedies do you favor?

In the late 1800s, a group of progressives in Philadelphia hired W. E .B. Du Bois, then a young African American scholar, to study the Seventh Ward in Philadelphia. To these progressives, poverty was a virus that needed to be contained before it contaminated "the closely allied product just outside the almshouse door." They wanted DuBois to diagnose the exact nature of the "virus" among the city's African Americans. They told him: "We want to know precisely how this class of people live; what occupations they follow; from what occupations are they excluded; how many of their children go to school; and to ascertain every fact which will throw light on this social problem."[6] Du Bois took the job because he believed that "the world was thinking wrong about race, because it did not know."[7] He was the convinced "the facts" would reveal the truth. How is his view of the power of "the facts" similar to the views of Riis and Addams? To your own views? Find out more about Du Bois's study of African Americans in Philadelphia. To what extent did "the facts" he uncovered change the way people thought about race in the city?

How do Riis and Addams use facts—particularly statistics—to define a problem? To suggest remedies? What are the advantages of using numbers and other data to document a problem? Does agreement on the nature of the problem necessarily mean agreement on a solution? How do you account for the differences in the remedies Riis and Addams suggest? Compare and contrast their use of statistics to that of Charles Davenport and Henry Goddard (Chapter 3). What similarities do you notice? How do you account for differences?

How does Addams define the word *citizen*? What does she see as the duties of a citizen? What rights does she seem to think every citizen enjoys? Addams founded Hull House at a time when women had the right to vote only in Wyoming. In fact, most married women in 1889 did not even have the right to the property they brought to their marriage or the money they earned on the job. In most states, both belonged to their husbands. Yet even as women struggled to expand their citizenship rights, some like Jane Addams took an active role in the political issues of their day. In doing so, how do they expand our understanding of what it means to be a citizen in a democracy? Of the ways an individual can make a positive difference in the world?

---

1. *The Battle of the Slum* by Jacob A. Riis, 1902.
2. *How the Other Half Lives* by Jacob A. Riis, 1890. Republished by Hill and Wang, 1957, p. 1.
3. Ibid., p. 3.
4. *Twenty Years at Hull House* by Jane Addams. First published in 1910 and reissued by Signet and New American Library.
5. *What Eight Million Women Want* by Rheta Childe Dorr. Boston, 1910, p. 327.
6. Quoted in *W. E. B. Du Bois: Biography of a Race, 1868-1919* by David Levering Lewis. Henry Holt, 1993, p. 188.
7. Ibid., p. 189.

Reading 9

Jane Addams and other progressives focused on issues like regular garbage collection, clean streets, and waste removal for an important reason—fear of epidemics of cholera, typhus, and other diseases associated with crowding, poor sanitation, and filth. Almost every city had experienced such outbreaks in the 19th century. As a result, many Americans regarded cities as dangerous places to live. Because epidemics often began in a city's poorest neighborhoods, many held the residents of those neighborhoods—especially immigrants—responsible.

Quoting New York health workers, reporter Jacob R. Riis labeled the Lower East Side, then a predominately Jewish neighborhood, as "the typhus ward" in *How the Other Half Lives.* He described it as a place where diseases "sprout naturally among the hordes that bring the germs with them from across the sea and whose instinct is to hide their sickness lest the authorities carry them off to be slaughtered."[1]

As early as 1862, Arthur B. Stout, a physician, expressed a similar view of Chinese immigrants in a report entitled "Chinese Immigration and the Physiological Causes of the Decay of a Nation."[2] After reading it, the California Board of Health asked him to investigate the harm to San Francisco that results from "the combined intermixture of races and the introduction of habits and customs of a sensual and depraved people in our midst . . . with hereditary vices and engrafted peculiarities."[3] Stout's report confirmed their fears.

Similar fears led Congress to pass the Chinese Exclusion Act in 1882. It was the first law to single out the residents of a single nation as "unsuitable" for residence in the United States. In the years that followed, Chinese already in the United States were repeatedly blamed for various epidemics, often with dubious evidence to support the claim or none at all. Increasingly, health officials in San Francisco and elsewhere came to see them as a "laboratory of infection" in the heart of the city "distilling its deadly poison by day and by night and sending it forth to contaminate the atmosphere of the streets and houses of a populous, wealthy, and intelligent community."[4] Then on February 1, 1900, the following story appeared in newspapers across the country:

> The steamship Australia . . . from Honolulu, arrived [in San Francisco] today and reports that up to the time of her departure forty-one deaths from the plague had occurred and there was a total of fifty-two cases.

In an effort to stamp out the plague, it was decided to burn one of the blocks in Chinatown [in Honolulu]. The fire was started and it gained such headway that the fire department could not control it. The flames spread rapidly from one block to another and soon the whole Chinese quarter was in flames. Hardly a house was left standing; 4,500 people were rendered homeless and they are now living in tents.[5]

In San Francisco, the largest port on the West Coast, the story created an uproar. As confused and often contradictory rumors of plague spread through the city, officials confirmed three cases of bubonic plague in Honolulu. The plague was as terrifying in 1900 as it was during the Middle Ages. Almost always fatal, it had no known cure and victims suffered agonizing deaths. Today scientists believe that fleas from infected rats carried the plague bacillus onto virtually every ocean-going ship at the turn of the 20th century. In 1900, people knew only that the disease was associated with filth and famine.

Public health officials in Hawaii tried to reassure Americans by announcing that they had all the chemicals needed for "proper destruction of the microbes." To underscore that message, the San Francisco *Examiner* published a detailed account of the sterilization of a steamship with "formaldehyde gas" before it was allowed to leave Honolulu.[6] The story quieted fears for a time.

Then on March 6, the body of Chick Gin, a storekeeper, was found in the basement of a hotel in San Francisco's Chinatown. At the time, city health officials required an examination into the death of any Chinese who was not under the care of a "Caucasian" physician. So officials collected tissue samples from the corpse. Before they could begin to analyze them, panicky health officers ordered the police to evacuate all "Caucasians" from Chinatown and then cordon off the community. On March 7, 25,000 Chinese residents awoke to find themselves separated from "white" neighbors by ropes that looped around a 14-block area.

Five days later, officials revealed that Gin had died of bubonic plague. At first no new cases were reported and people began to relax. Then suddenly, there were three more suspicious deaths. In the weeks that followed, quarantines were imposed on Chinatown and then lifted. There were rumors of missing corpses, stolen tissue samples, and hidden information. In June, after health officers documented ten cases in the Chinese quarter, the city sent 75 inspectors and 50 policemen to search every building in Chinatown and root out every ailing resident. City workers even built fumigating stations at the edge of Chinatown so that "white" San Franciscans who worked in the area could be disinfected before they returned home each evening.

The Chinese were then "absolutely shut away from the rest of the world." Streetcars did not enter Chinatown nor was the mail delivered as sixty policemen stood guard. Fear fed on fear. As early as March, *Organized Labor*, a union publication, warned, "The almond-eyed Mongolian is waiting for his opportunity, waiting to assassinate you and your children with one of his many maladies."[7]

The Chinese had fears too. There was talk of shipping them to an isolated island in San Francisco Bay and then burning their homes and businesses. One local newspaper demanded, "Clear the foul spot from San Francisco and the debris to the flames." Fearful that city officials were also planning to poison their water supply, Chinese leaders placed guards around the water tanks. When a white sanitation worker who was new to the city wandered near one of those tanks, he was almost killed.

Again and again, the Chinese demanded that the quarantine be lifted. When their protests went unheard, they turned to the courts for help. The first case focused on an attempt to forcibly inoculate the Chinese with an experimental drug believed to prevent the plague. Lawyers for Wong Wai, a Chinese merchant, argued that forced inoculation violated his right to pursue a lawful business and denied him "equal protection of the laws." Judge William Morrow agreed. He ruled that the measures the city adopted were "not based upon any established distinction in the conditions that are supposed to attend the plague or the persons exposed to the contagion." Instead, he argued, officials took measures that were "boldly directed against the Asiatic or Mongolian race as a class without regard to the previous condition, habits, exposure or disease, or resident of the individual on the unproven assumption that this race is more liable to the plague than any other."[8]

The second suit filed by the Chinese focused on the legality of the quarantines. This time a grocer in Chinatown, Jew Ho, filed the complaint on behalf of other residents. Ho challenged the quarantine as arbitrary and discriminatory. His lawyers argued that while white San Franciscans were allowed to enter and leave Chinatown as they pleased, Chinese residents were effectively under house arrest. They also noted that despite its claims that the quarantine was necessary, the Board of Health had made no provision to feed or care for isolated members of the Chinese community. The court agreed with Ho but did permit the city to quarantine specific buildings that officials believed were contaminated.
The plague did not end with a court order. It continued to claim lives at the rate of about one victim every two weeks. On August 11, the first "Caucasian" in the city died of the disease. By 1904, officials had documented 121 cases and 112 deaths. The vast majority of the victims were Chinese.

CONNECTIONS

Many stories have a moral or lesson. What is the moral of the story of the plague? How does it deepen our understanding of what it means to be outside a community's universe of obligation?

What is the power of fear? How does it thrive on rumors, myth, and misinformation? Why does fear often lead to violence?

Congress passed the Chinese Exclusion Act in 1882, partly out of fear of epidemics. In the years that followed, violence against people of Chinese ancestry increased dramatically. What is the connection between racism and violence?

What aspects of the work of Charles Davenport and other eugenicists might appeal to San Franciscans and other Americans who were frightened by the plague? By fears of contamination?

Howard Markel is a physician who has written a book about the experiences of Eastern European Jewish immigrants during typhus and cholera epidemics in New York City in 1892. In the concluding chapter, he notes, "The microbe as an agent of illness and death is the ultimate social leveler. It binds us and, when transmitted through a filter of fear, has the potential to divide."[9] In what sense is a microbe a "social lever"? How do Markel's comments about microbes apply to the experiences of Chinese Americans in San Francisco in the early 1900s? To victims of the AIDS epidemic in recent years?

---

1. *How the Other Half Lives* by Jacob A. Riis, 1890. Republished by Hill and Wang, 1957, p. 1.
2. *Silent Travelers: Germs, Genes, and the "Immigrant Menace"* by Alan M. Kraut. HarperCollins Publishers, 1994, p. 80.
3. Ibid., p. 81.
4. Ibid., p. 82.
5. *Washington Post,* February 1, 1900.
6. This account is based in part on pages 164-166 in *America 1900* by Judy Crichton. Henry Holt and Company, 1998.
7. Quoted in *America in 1900* by Noel Jacob Kent. M.E. Sharpe, 2000, p.107.
8. *Silent Travelers: Germs, Genes, and the "Immigrant Menace"* by Alan M. Kraut. HarperCollins Publishers, 1994, pp. 91-92.
9. *Quarantine!* by Howard Markel. Johns Hopkins University Press, 1997, p. 192.

Reading 10

Progressive reformers were primarily middle-class white Americans who were uncomfortable with many of the changes that were taking place in American life at the turn of the 20th century. They viewed immigrants, African Americans, working families, and the poor, as groups in need of help and advice rather than as independent individuals with voices and ideas of their own. African Americans like Ida B. Wells challenged those views. So did many immigrants. They did not see themselves as problems but as ordinary people who wanted many of the things that other Americans wanted—a safe place to live, a good job, and opportunities for themselves and their children. To achieve these goals they confronted and sometimes overcame extraordinary obstacles. Pauline Newman's story challenges the stereotypes that shaped the way many Americans viewed the nation's newest arrivals.

Newman and her family came to New York City from Lithuania, a country in Eastern Europe, in 1901. Although she was only about eight years old at the time, within weeks of her arrival she was working in a factory that made shirtwaists—linen dresses popular with many women at the turn of the 20th century. In 1975, she told a group of young people:

> I'd like to tell you about the kind of world we lived in 75 years ago because all of you probably weren't even born then. . . . That world 75 years ago was a world of incredible exploitation of men, women, and children. I went to work for the Triangle Shirtwaist Company in 1901. The corner of a shop would resemble a kindergarten because we were young, eight, nine, ten years old. It was a world of greed; the human being didn't mean anything. The hours were from 7:30 in the morning to 6:30 at night when it wasn't busy. When the season was on we worked until 9 o'clock. No overtime pay, not even supper money. There was a bakery in the garment center that produced little apple pies the size of this ashtray [holding up an ashtray for group to see] and that was what we got for our overtime instead of money.
>
> My wages as a youngster were $1.50 for a seven-day week. I know it sounds exaggerated but it isn't; it's true. . . . I worked on the 9th floor with a lot of youngsters like myself. When the operators were through with sewing shirtwaists, there was a little thread left, and we youngsters would get a little scissors and trim the threads off.

And when the inspectors came around, do you know what happened? The supervisors made all the children climb into one of those crates that they ship material in, and they covered us over with finished shirtwaists until the inspectors had left, because of course we were too young to be working in the factory legally.

The Triangle Waist Company was a family affair, all relatives of the owner running the place, watching to see that you did your work, watching when you went into the toilet. And if you were two or three minutes longer than foremen or foreladies thought you should be, it was deducted from your pay. If you came five minutes late in the morning because the freight elevator didn't come down to take you up in time, you were sent home for a half a day without pay.

. . . The early sweatshops were usually so dark that gas jets [for light] burned day and night. There was no insulation in the winter, only a pot-bellied stove in the middle of the factory. . . . Of course in summer you suffocated with practically no ventilation. There was no drinking water, maybe a tap in the hall, warm, dirty. What were you going to do? Drink this water or none at all.

The conditions were no better and no worse than the tenements where we lived. You got out of the workshop, dark and cold in winter, hot in summer, dirty unswept floors, no ventilation, and you would go home. What kind of home did you go to? Some of the rooms didn't have any windows. I lived in a two-room tenement with my mother and two sisters and the bedroom had no windows, the facilities were down in the yard, but that's the way it was in the factories too.

We wore cheap clothes, lived in cheap tenements, ate cheap food. There was nothing to look forward to, nothing to expect the next day to be better. Someone asked me once: "How did you survive?" And I told him, "What alternative did we have? You stayed and you survived, that's all."[1]

Newman, however, did more than stay and survive. In an interview with author Joan Morrison, she described her efforts to get an education:

At first I tried to get somebody who could teach me English in the evening, but that didn't work out because I don't think he was a very good teacher, and, anyhow, the overtime interfered with private lessons. But I mingled with people. I joined the Socialist Literary Society. Young as I was and not very able to express myself, I decided that it wouldn't hurt if I listened. There was a Dr. Newman, no

> relation of mine, who was teaching at City College. He would come down to the Literary Society twice a week and teach us literature, English literature. He was very helpful. He gave me a list of books to read, and as I said, if there is a will you can learn. We read Dickens, George Eliot, the poets. I remember when we first heard Thomas Hood's "Song of the Shirt." I figured that it was written for us. You know, because it told of the long hours of "Stitch! Stitch! Stitch!" I remember one of the girls said, "He didn't know us, did he?" And I said, "No, he didn't." But it had an impact on us. . . .
>
> I regretted that I couldn't go even to evening school, let alone going to day school, but it didn't prevent me from trying to learn and it doesn't have to prevent anybody who wants to. I was then and still am an avid reader. Even if I didn't go to school I think I can hold my own with anyone, as far as literature is concerned.
>
> Conditions were dreadful in those days. We didn't have anything. If the season was over, we were told, "You're laid off. Shift for yourself." How did you live? After all, you didn't earn enough to save any money. Well, the butcher trusted you. He knew you'd pay him when you started work again. Your landlord, he couldn't do anything but wait, you know. Sometimes relatives helped out. There was no welfare, no pension, no unemployment insurance. There was nothing. . . .
>
> But despite that, we had good times. In the summer we'd go to Central Park and stay out and watch the moon rise; go to the Palisades and spend the day. We went to meetings, too, of course. We had friends and we enjoyed what we were doing. We had picnics. And, remember, in that time you could go and hear [tenor Enrico] Causo for twenty-five cents. . . . Of course we went upstairs [to the balcony], but we heard the greatest soloists, all for a quarter, and we enjoyed it immensely. We loved it. We'd go Saturday night and stand in line no matter what the weather. In the winter we'd bring blankets along. Just imagine, the greatest artists in the world, from here and abroad, available to you for twenty-five cents.[2]

By the time she was 15, Newman was not only reading poetry and attending concerts but also organizing a labor union at the Triangle Shirtwaist Company. By 1909, she was working full-time as a union organizer. That year, as a result of her efforts and those of other organizers, thousands of garment workers in New York City went on strike for higher wages, a shorter work week, and safer working conditions. Newman recalled the mood of the workers the day the strike began:

> Thousands upon thousands left the factories from every side, all of them walking down toward Union Square. It was November, the cold winter was just around the corner. . . .
>
> I can see the young people, mostly women, walking down and not caring what might happen. The spirit, I think, the spirit of a conqueror led them on. They didn't know what was in store for them, didn't really think of the hunger, cold, loneliness, and what could happen to them. They just didn't care on that particular day; that was their day.[3]

In the days that followed, the women quickly learned that a strike required more than "spirit." After visiting their union hall, a reporter for the *New York Sun* wrote: "There, for the first time in my comfortably sheltered, Upper West Side life, I saw real hunger on the faces of my fellow Americans in the richest city in the world."[4] The young strikers also faced arrest.

Picketing—carrying signs and banners outside a place of employment to express grievances and keep strikebreakers out—was illegal. Newman recalled, "The judge, when one of our girls came before him, said to her: 'You're not striking against your employer, you know, young lady. You're striking against God,'" and sentenced her to two weeks on Blackwell's Island, which is now Welfare Island. And a lot of them got a taste of the club."[5]

Despite hunger and the threat of jail, the union enrolled a thousand new members each day. Some estimate that as many as 20,000 men, women, and children participated in the strike. As money began to run out, leaders sent organizers like Pauline Newman to other cities to seek help from women's clubs and other unions. They also won the support of prominent New York women and settlement house leaders like Lillian Wald and Mary Simkovitch, who used their connections to protect the strikers, raise money, and press factory owners to settle with workers.

The strike lasted three months. It officially ended on February 15, 1910. Historians Irving Howe and Kenneth Libo have described the strike as "an uprising of people who discovered on the picket lines their sense of dignity and self. New emotions swept the East Side, new perceptions of what immigrants could do, even girls until yesterday mute. '*Unzere vunderbare farbrente meydlekh*' (our wonderful, fervent girls) an old-timer called them."[6]

Newman was less positive. She told an interviewer, "We didn't gain very much at the end of the strike. I think the hours were reduced to fifty-six a week or something like that. We got a ten percent increase in wages. I think that the best thing that the strike did was to lay a foundation to build a union."[7] Newman's enthusiasm was tempered by the realization that many employers, including the owners of the Triangle Shirtwaist Company where she had once worked, refused to negotiate with the union. They simply fired the strikers and hired replacements.

One year later, on March 25, 1911, a fire broke out at the Triangle Shirtwaist Company. The fire claimed the lives of 146 workers—143 of them were women and children who worked on the ninth floor. One former striker noted, "If the union had won, we would have been safe. Two of our demands were for adequate fire escapes and for open doors from the factories to the street. But the bosses defeated us and we didn't get the open doors or the better fire escapes. So our friends are dead."[8]

After the fire, Newman and other union organizers persuaded lawmakers and reformers to pass what Frances Perkins hailed as "the greatest battery of bills to prevent disaster and hardships." She believed that these new laws were the legacy of the victims in the Triangle Fire. Perkins was the chief investigator for a state commission that investigated factory conditions after the fire. In her view those laws "in some way" "paid the debt society owed to those children, those young people who lost their lives in the Triangle Fire. It's their contribution to the people of New York that we have this really magnificent series of legislative acts to protect and improve the administration of the law regarding the protection of work people in the City of— in the State of New York."[9]

CONNECTIONS

Pauline Newman never uses the words *progress* or *civilized*. How are those ideas reflected in her account? In reflecting on the world of 1900, Pauline Newman compares that world with life in the United States in 1975. How do those changes compare to the ones that took place in Walt Whitman's lifetime (Reading 1)?

Dictionaries define a *union* as a group of people who form an organization to work for a common cause or interest. What, then, is a *labor union*? What is its common cause or interest? What is a strike? Why has it become an important tool for labor unions?

After the strike, Pauline Newman recalls that she and her co-workers tried to educate themselves by reciting English poetry to improve their understanding of the language.[10] One of their favorites was Percy Bysshe Shelley's "Mask of Anarchy"—particularly the last two stanzas. The poem was written to commemorate a mass meeting of British reformers who were attacked by soldiers on August 16, 1819. Several were killed and hundreds were injured.

> And these words shall then become
> Like Oppression's thundered doom
> Ringing through each heart and brain,
> Heard again-again-again!
>
> Rise like lions after slumber
> In unvanquishable number!
> Shake your chains to earth, like dew
> Which in sleep had fallen on you—
> Ye are many, they are few.

What is the lesson of the poem? Why might it have special appeal to young immigrant workers?

Look carefully at the Bill of Rights—the first ten amendments to the Constitution. How did workers use those rights to fight injustices? How is their use of those amendments similar to the way Chinese immigrants used those same rights? How do you account for differences?

How does Newman's account of her early years in the factory challenge stereotypes about immigrants held by Charles Davenport (Chapter 3) and other eugenicists? How does her account challenge stereotypes about women and their role in society? Why might her activities have seemed threatening to some progressive reformers?

What do Howe and Libo mean when they write that the strikers "discovered on the picket lines their sense of dignity and self"? Would eugenicists like Charles Davenport agree? How do you think he might have described the young strikers?

The fire was a turning point in the lives of many people, including Frances Perkins, then a young social worker who saw the tragedy with her own eyes. In her role as chief investigator for the state commission, she set out to educate lawmakers. She writes in her memoirs:

> We used to make it our business to take Al Smith [then a state lawmaker] . . . to see the women, thousands of them, coming off the ten-hour night-shift on the rope walks in Auburn [New York]. We

made sure that Robert Wagner [also a lawmaker] personally crawled through the tiny hole in the wall that gave egress to a steep iron ladder covered with ice and ending twelve feet from the ground, which was euphemistically labeled "Fire Escape" in many factories. We saw to it that the austere legislative members of the Commission got up at dawn and drove with us for an unannounced visit to a Cattaraugus County cannery and that they saw with their own eyes the little children, not adolescents, but five-, six-, and seven-year-olds, snipping beans and shelling peas. We made sure that they saw the machinery that would scalp a girl or cut off a man's arm. Hours so long that both men and women were depleted and exhausted became realities to them through seeing for themselves the dirty little factories.[11]

How is the kind of education Perkins provided lawmakers different from the kind Jacob Riis provides in *How the Other Half Lives*? Why do you think she placed such importance on lawmakers meeting workers and seeing their conditions rather than reading about them in a book or a report? What did she want lawmakers to learn? What did she hope they would remember?

---

1. From a taped talk by Pauline Newman to union women from Trade Union Women's Studies, a program of the New York State School of Industrial Labor Relations, Cornell University, March, 1975. Quoted in *We Were There: The Story of Working Women in America* by Barbara Mayer Wertheimer. Pantheon Books, 1977, p. 294-295.
2. Interview in *American Mosaic* by Joan Morrison and Charlotte Fox Zabusky. New American Library, 1980, pp. 11-12.
3. From a taped talk by Pauline Newman, March, 1975. Quoted in *We Were There: The Story of Working Women in America* by Barbara Mayer Wertheimer. Pantheon Books, 1977, p. 301.
4. Quoted in *How We Lived* by Irving Howe and Kenneth Libo. Richard Marek Publishers, 1979, p. 182.
5. Interview in *American Mosaic* by Joan Morrison and Charlotte Fox Zabusky. New American Library, 1980, p. 13.
6. *How We Lived* by Irving Howe and Kenneth Libo. Richard Marek Publishers, 1979, p. 182.
7. Interiew in *American Mosaic* by Joan Morrison and Charlotte Fox Zabusky. New American Library, 1980, pp. 12-13.
8. Quoted in *The Triangle Fire* by Leon Stein. J.B. Lippincott, 1962, p. 168.
9. *Lectures of Frances Perkins*, Collection /3047, 30 September 1964, Cornell University, Kheel Center for Labor-Management Documentation and Archives, Ithaca, NY.
10. Quoted in *We Were There: The Story of Working Women in America* by Barbara Mayer Wertheimer. Pantheon Books, 1977, p. 308.
11. *The Roosevelt I Knew* by Frances Perkins. Viking Press, 1946, pp. 17, 22.

5. Eugenics and the Power of Testing

Most of us are wholly convinced that the future of mankind depends in no small measure upon the development of the various biological and social sciences.

Robert Yerkes

Chapter 4 explores the historical context in which eugenics flourished by examining the ways Americans in the late 1800s and early 1900s answered the question: What do you do with a difference? Chapters 5, 6, and 7 consider the impact of eugenics on public policy at the local, state, and national levels.

Some eugenicists sought to protect the nation from the danger of "inferior genes" by encouraging "good families" to have as many children as possible. Others favored "negative eugenics"—keeping the "unfit" from breeding, with force if necessary. Both approaches required an efficient way of determining who was "fit" and who was not. The key to protecting the nation's gene pool lay in finding a method for measuring intellectual ability.

Eugenicists believed that a French diagnostic test developed in 1905 provided the tool they needed to separate the "fit" from the "unfit." They called it an "intelligence test" even though it was originally developed to predict how children would do in school and which of them might need extra help. Among the few to suggest that new test was based less on science than on a "will to believe" was journalist Walter Lippmann. In the first of a series of articles in the *New Republic*, he wrote:

> Without offering any data on all that occurs between conception and the age of kindergarten, they announce on the basis of what they have got out of a few thousand questionnaires that they are measuring the hereditary mental endowment of human beings. Obviously this is not a conclusion obtained by research. It is a conclusion planted by the will to believe.[1]

Despite such criticism, eugenicists convinced many educators, religious leaders, politicians, and ordinary citizens that intelligence testing could not only improve education but also end poverty, prevent crime, and wipe out disease by identifying the individuals responsible for these problems. In an age dazzled by scientific and mechanical wonders, few were willing to criticize a seemingly scientific theory. Indeed, many saw "men of science" as above the rough and tumble of politics.

The readings in Chapter 5 raise troubling questions about the power of tests not only to categorize and rank individuals and groups but also limit their possibilities. The chapter also reveals how science can be twisted to justify social inequalities, deny opportunities, and legitimize discrimination. British scientist P. B. Medawar has described the scientific method as taking "for granted that we guess less often right than wrong, but at the same time ensures that we need not persist in error if we earnestly and honestly endeavor not to do so."[2] Yet long after Thomas Hunt Morgan and other scientists had shown that the laws of heredity are more complicated than "breeding the best with the best," eugenicists were still trying to segregate "mental defectives." Long after Franz Boas and other anthropologists had shown that intelligence is shaped at least in part by culture and environment, eugenicists were still seeking ways to "protect" the "superiority of the white race" by outlawing interracial marriages. Chapter 5, along with chapters 6 and 7, considers not only why eugenicists "persisted in error" but also the consequences of those errors on public policy long ago and today.

---

1. Quoted in *The Mismeasure of Man* by Stephen Jay Gould. W.W. Norton, 1991, p. 174.
2. *The Limits of Science* by P. B. Medawar. Harper & Row, 1984, p. 101.

Science, Eugenics, and Propaganda

Reading 1

The word *science* comes from *scientia*, the Latin word for knowledge. British scientist P. B. Medawar, a Nobel laureate, once described the term as "knowledge hard won, in which we have much more confidence than we have in opinion, heresay, and belief." In response to those who argued, "Unless it's successful, you don't call it science," he wrote:

> What rot! I have been engaged in scientific research for about fifty years and I rate it highly scientific even though very many of my hypotheses have turned out mistaken or incomplete. This is our common lot. It is a layman's illusion that in science we caper from pinnacle to pinnacle of achievement and that we exercise a Method which preserves us from error.[1]

Like Medawar, most scientists believe that research must be open to criticism, revision, and debate because any hypothesis may be "mistaken or incomplete." Eugenicists took a different approach to research. They used it to confirm and disseminate what they already believed. The result is propaganda.

Propaganda is often defined as the dissemination of information for the purpose of persuasion or to advocate a particular agenda. Those who create propaganda seldom want careful scrutiny or criticism. Their goal is to bring about a specific action. Eugenicists organized fairs and exhibitions to promote their ideas and detailed them in books, magazines, and newspapers. Ministers preached eugenics from the pulpit and teachers incorporated it into their lessons. Eugenicists supplied civic groups, social clubs, and libraries with speakers and free study materials. They also arranged a variety of contests to introduce Americans to the principles of eugenics—including the idea that intelligence is shaped almost solely by heredity and is linked to morality. Among the most popular of these contests were the Fitter Families competitions. The first was held at a state fair in Topeka, Kansas, in 1920. By the end of the decade, they were featured, along with eugenic exhibits, at fairs in Kansas and in a number of other states. Historian Daniel J. Kevles says of these contests:

> At state fairs, the Fitter Families were held in the "human stock" sections. ("The time has come," a contest brochure explained, "when the science of human husbandry must be developed, based on the principles now followed by scientific agriculture, if the better elements of our civilization are to dominate or even survive.") Any healthy family could enter. Contestants had only to provide an examiner with

> the family's eugenic history. . . . At the 1924 Kansas Free Fair, winning families in three categories—small, average, and large—were awarded a Governor's Fitter Family Trophy, presented by Governor Jonathan Davis. "Grade A Individuals" won a Capper Medal, named for United States Senator Arthur Capper and portraying two diaphanously garbed parents, their arms outstretched toward their (presumably) eugenically meritorious infant. A fair brochure noted that "this trophy and medal are worth more than livestock sweepstakes or a Kansas oil well. For health is wealth and a sound mind in a sound body is the most priceless of human possessions."[2]

Eugenicists also offered prizes to the "best baby" and young couples about to embark on a "eugenic marriage." School children were ranked not only according to their intelligence but also their mental outlook, height, dental hygiene, vision, and hearing. For example, a child whose height deviated in either direction from the Hastings' Age-Height Tables, which stated the "normal height" for a child at a particular age, received a low score.

First prize winners in a "Fitter Family Contest" Topeka, KS, 1927.

The eugenics exhibits at these fairs often featured billboards like the one shown in the photograph on page 144. The lights flashed every 15 seconds to indicate how often $100 of the taxpayers' money went for the care of a mentally deficient person born in the United States. Other lights flashed every seven and a half minutes to indicate how often a "high grade" person was born. "How long are we Americans to be so careful about the pedigree of our pigs and chickens and cattle—and then leave the ancestry of our children to chance or to blind sentiment?" asked a nearby sign. A pamphlet published in 1915 by the Juvenile Protective Association of Cincinnati reveals yet another way eugenicists tried to alert Americans to the "menace of the feebleminded." (See cover, page 145.)

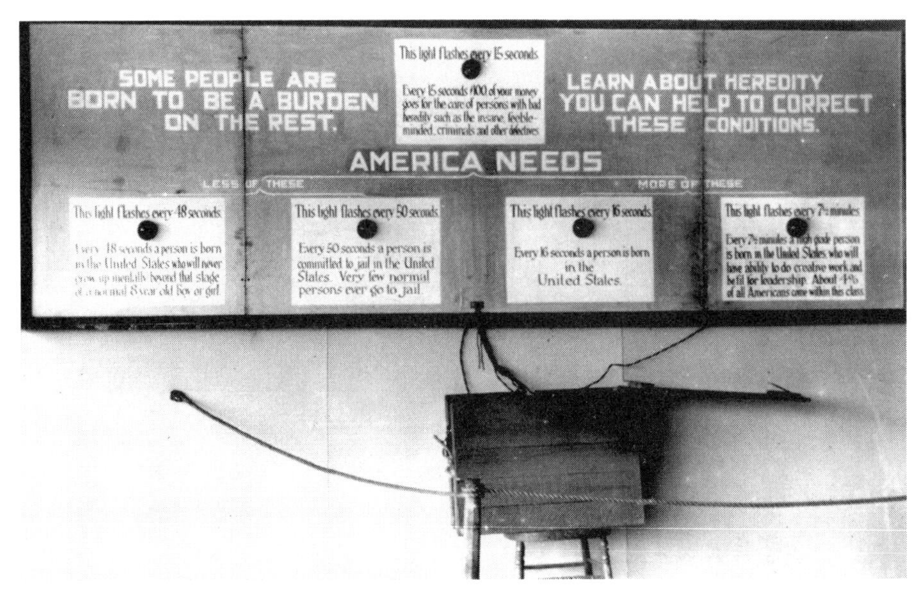

Exhibit at a eugenics fair.

CONNECTIONS

The contests, the pamphlet cover on page 145, and the exhibit shown above are all examples of propaganda—attempts to use emotion to sway public opinion. What feelings does each evoke? What techniques does each use to promote those feelings? To alter or adjust perceptions? For example, what is the message or moral of the various eugenics contests? At whom is that message aimed? To what emotions do the contests appeal? How do you think the winners of these contests regarded themselves? Their lower-ranked neighbors? How do you think the losers saw themselves and others?

Working in small groups, take a closer look at the cover of the pamphlet on the following page:
—What do you see? Try not to explain the drawing, simply describe what you notice. Have someone in the group record your observations and those of your classmates. You may also want to chronicle your impressions in your journal.
—Interpret the drawing. Why do you think the artist placed a man's face at the center of the wheel? How does it reinforce the words on the diagram? What message do the words and wheel convey? Would the message be different if the figure at the center of the wheel were an elderly woman? An African American? A parent and child?
—What is the significance of the spokes that emanate from the man's face to the outer ring? How does this technique reinforce the artist's message? At whom is the message aimed?

—What characteristics make the drawing seem scientific? Authoritative?
—Identify the emotions that the drawing evokes in you and others in your group. What might have been the reaction of a person seeing this image in 1915?

Study the traveling exhibition shown in the photograph on page 144 much the way you studied the pamphlet cover. Keep in mind that the exhibition, unlike the drawing, was three-dimensional. The flashing lights were designed to turn a viewer's attention to the short messages that appeared on the various posters in the exhibit. What effect might those lights have on a viewer?

In the introduction to this chapter, Walter Lippmann compared conclusions based on scientific research to those "planted by the will to believe." To what extent are the images in this reading based on scientific research? "Planted by the will to believe"? What similarities do you see in the messages each image conveys? How do you account for differences? Which seems more scientific?

The two images included in this reading describe a problem but offer no solution. What solutions is a person likely to suggest after viewing them? After participating in a "Fitter Family" contest? Compare those solutions with the one Charles Davenport offers in *Heredity in Relation to Eugenics* (pages 75-76). What similarities do you notice? What differences are most striking?

This reading describes how eugenicists in the early 1900s tried to communicate their ideas to a broad, general audience. How might a group today popularize an idea? What technologies might they use? What methods do you think would be most effective? Least effective? Be prepared to state why you have chosen a particular strategy. How might those who disagree with an idea get heard?

For more information on the "Fitter Family contests" and eugenic displays at state fairs, visit a website devoted to the archives of the Eugenics Record Office at Cold Spring Harbor: *www.eugenicsarchive.org/eugenics.*

---

1. *The Limits of Science* by P. B. Medawar. Harper & Row, 1984, p. 101.
2. *In the Name of Eugenics* by Daniel J. Kevles. Harvard University Press, 1985, 1995, p. 62.

Reading 2

In his textbook, *Heredity in Relation to Eugenics*, Charles Davenport argued, "It is just as sensible to imprison a person for feeble-mindedness or insanity as it is to imprison criminals belonging to such strains. The question of whether a given person is a case for the penitentiary or the hospital is not primarily a legal question but one for a physician with the aid of studies of heredity and family histories." Throughout the early 1900s, Davenport and other eugenicists repeatedly warned the nation of the threat posed by the "unfit"—the so-called "menace of the feebleminded."

Caretakers at institutions for people with mental disabilities popularized the term *feebleminded* in the late 1800s. Although they never clearly defined it, the word originally referred to an individual who was not only "hereditarily deficient in mental capacity" but also a "burden" to society. By the turn of the century, the word had a new connotation—the "feebleminded" were more than a "burden," they had become a "threat" to society. Lewis Terman, a noted psychologist and eugenicist, explained:

> Not all criminals are feebleminded, but all feebleminded persons are at least potential criminals. That every feebleminded woman is a potential prostitute would hardly be disputed by anyone. Moral judgment, like business judgment, social judgment, or any other kind of higher thought process, is a function of intelligence. Morality cannot flower and fruit if intelligence remains infantile.[1]

The campaign against the "feebleminded" had consequences. Lawmakers in state after state responded by building special institutions to separate the "feebleminded" from other Americans. By 1917, 31 of the nation's 48 states supported "homes," "colonies," or "schools" for mentally retarded and epileptic persons (regardless of intelligence).[2]

The campaign also affected how the "menace" was defined. In 1920, a writer for *Mental Hygiene,* a professional journal, explained, "Whereas ten years ago 80% of [admissions] were idiots and imbeciles and only 20% border-line cases or morons, now 20% are of the idiot and imbecile class and 80% are morons or border-line cases."

The vast majority of those admitted to institutions for the "feebleminded" in the early 1900s shared other characteristics as well. Almost all of them were white. There were no comparable institutions for African Americans at the time.

Almost all of the inmates were poor and the vast majority were female. In many respects, "Deborah Kallikak" (pages 82-84) was a typical inmate. An article in a professional journal reflected the "conventional wisdom":

> Feeble-minded women are almost invariably immoral, and if at large usually become carriers of venereal disease or give birth to children who are as defective as themselves. The feeble-minded woman who marries is twice as prolific as the normal woman.
>
> There is no class of persons in our whole population who, unit for unit, are so dangerous or so expensive to the state. This excepts no class, not even the violently insane. There are much more dangerous and expensive than the ordinary insane or the ordinary feeble-minded or the ordinary male criminal. Why is this? They are dangerous because being irresponsible wholly or in part they become the prey of the lower class of vile men and are the most fertile source for the spread of all forms of venereal disease. They have not the sense or the understanding to avoid disease or any care as to its spread. They are most expensive to the state because they are the most fruitful source of disease and mentally defective children who are apt to become state charges.[3]

These assumptions and beliefs shaped both public policy and private actions. Until the 20th century, all but the most severely retarded lived much as their neighbors did. They attended the same schools, prayed in the same churches and synagogues, paid the same taxes, and worked at many of the same jobs. They, too, married and had children. By the early 1900s, eugenic propaganda had persuaded a growing number of Americans that the "feebleminded" should not only be separated from the rest of society but also denied the rights that other Americans enjoyed.

In 1907, Congress closed the nation's borders to immigrants who were "feebleminded." A few years later, nine states had laws banning the sale of alcohol to such individuals and one forbade the sale of firearms. By the 1920s, 39 states denied the "feebleminded" the right to marry. In 18 states they could not vote, and six states denied them the right to enter into a contract. In some states, they could not serve in the National Guard and there was talk of removing the "feebleminded" from the U.S. armed forces.[4]

What did the growing isolation mean to those who were labeled as "unfit"? How did their families respond to their incarceration? For the most part, their feelings and emotions have been lost to history. Stories like "Deborah Kallikak's" offer some clues. So does a letter written in 1902 by a resident of a facility for the "feebleminded":

My dear Father:

I wish you would leave me come home for my birthday which is not far off. It comes on the 25th of September, which is Thursday. There is one question I wish to ask you it is this: if I ask you to take me home, you say you haven't the money and I run away why you seem to have it to bring me back, and that is what puzzles me. I only wish I could spend just one month with you, I would be more than satisfied, and you know I have been here exactly 9 years and haven't been home in a decent way yet, and I guess I never will. If you can't give me a little change, I will have to make it myself. I will never show my face near home, and you can depend on it.

Your unthought of Son.[5]

An experiment in an institution in New York State also offers insights into the way some young Americans responded to the labels. Although the directors of most institutions supported eugenicists in their calls for lifelong segregation of the mentally retarded, Charles Bernstein was among the few to challenge that idea in the early 1900s. Convinced of the power of education to help the "feebleminded" become self-sufficient, he began to release inmates after offering them some training. In a monthly newspaper, he often printed letters from recently discharged inmates. In 1917, a former inmate wrote:

Just a few lines to let you know that I am still alive and in the best of health. I am now in the US Navy. I enlisted July 9th and I am now at the Training Station at Newport, R.I. and expect to leave here on the ship next week for France.

This is a fine place down here. There are about 10,000 boys down here. There isn't a chance to get lonesome. There are a lot of boys in your institution who I think if they were in the navy it would make a man of them.

I was considered feeble-minded once, but I was given the chance to prove I was not. I am now in a place where you have to have a strong mind and be quick witted. I am proud to say that I am just as good as any of them. The reason for me getting out of that I once got in is that I made a fool out of the ones that tried to make a fool out of me. You must remember me, the kind of a boy that I was, so if there are any others like me, give them a chance, they will make good.[6]

A few years later, yet another former inmate reported:

I have just received my report card Friday, so I thought I'd let you know my marks. Algebra, three; Civics, three; English, two; Latin, four; Gym, three, and Citizenship, two. On the back of the card it told what the marks stood for and I will copy it for you. Group one includes those whose work is of the highest excellency, a distinction reached by few in a class; group two those whose work while not perfect is still so excellent that it is decidedly above the average of good work.[7]

CONNECTIONS

How do you explain why people in one century accepted individuals with the characteristics of "feeblemindedness" and people in the next century isolated them? What fears prompted the change? What does the reading suggest about the consequences of neighbor turning against neighbor? Record your ideas in your notebook so that you can add to your ideas or revise them as you continue reading.

How do you account for the fact that the majority of inmates in "homes for the feebleminded" were white females from poor families? What do those facts suggest about the way Americans were defining their "universe of obligation" in the early 1900s? What attitudes and values were reflected in those definitions?

Why do you think that women labeled as "feebleminded" were considered a burden to society and more dangerous than "the violently insane" or the "ordinary male criminal"? What did they threaten? Whom did they threaten?

What do the words of inmates and former inmates suggest about what it meant to be labeled as "feebleminded"? How did that label shape their identity—their sense of who they were and what they might become? How might their voices have shaped public policies aimed at the "menace of the feebleminded"? What questions might their experiences have raised about the meanings people attach to differences? About the power of labels?

Walter Lippmann coined the word *stereotype* in the 1920s. He defined the term as a "picture in our heads." He thought of stereotypes as both positive and negative. Today, any kind of stereotype is considered offensive, because it applies a small kernel of truth about some people to an entire group. According to sociologist Herbert J. Gans, "Negative labels rarely stereotype only behavior; more often they transform and magnify it into a character failing. As a result, welfare recipients become defective personalities or deficient moral types; that they are also family members, churchgoers, or neighbors is immaterial. Indeed, one of the purposes of labels is to strip labeled persons of other qualities." Research the way a particular group—the mentally or physically disabled, the poor, African Americans, Chinese Americans, Latinos—is portrayed in the news, in movies, and on TV. Brainstorm a list of ways the stereotypes you and your classmates uncovered might be revised or abandoned.

---

1. *The Measurement of Intelligence* by Lewis Terman. Houghton Mifflin, 1916, p. 11.
2. Quoted in *A History of Mental Retardation* by R.C. Scheerenberger. Paul H. Brooks Publishing Co., 1983, p.158.
3. "High-Grade Mental Defectives" by W. Bullard. *Journal of Psycho-Asthenics*, 1909, p. 15.
4. Quoted in *A History of Mental Retardation* by R.C. Scheerenberger. Paul H. Brooks Publishing Co., 1983, p. 155.
5. Ibid., p. 159.
6. Quoted in *Inventing the Feeble Mind* by James W. Trent. University of California Press, 1994, p. 210.
7. Ibid., pp. 210–211.

Identifying the "Unfit"

Reading 3

In the early 1900s eugenicists needed a cheap and efficient method of identifying people they considered "unfit." On a trip to Europe, Henry Goddard, who directed a laboratory for the study of mental deficiency at the Vineland Training School for Feeble-minded Boys and Girls in New Jersey (Chapter 3), learned about a new test that would allow him and others to easily measure and then identify the "feebleminded." He translated it into English with a few minor changes, and then administered it to inmates at the school. He labeled those who scored 25 points or lower, "idiots," those who scored between 25 and 55, "imbeciles," and those between 55 and 75, "morons."

The test Goddard discovered was created in 1905 by Alfred Binet, the director of the Psychology Laboratory at the Sorbonne in Paris. Binet saw the test as a technique for predicting how children would do in elementary school. He wanted to alert teachers to students in need of extra help. So Binet asked children to perform tasks much like the ones they would be expected to perform at school. As he noted, "One might almost say, 'it matters very little what the tasks are so long as they are numerous.'"

Binet and his colleague, Théodore Simon, compiled a long list of tasks that children between the ages of three and twelve were typically assigned in school. They placed an age level on each task based on what they thought was the youngest age at which a child could successfully perform it. Those tasks formed the basis of the Binet-Simon scale. Binet believed that the scale was simply a measure of a child's ability to perform specific tasks at a particular moment in the youngster's life. He warned against attaching greater meaning to the results:

> Some recent thinkers seem to have given their moral support to these deplorable verdicts by affirming that an individual's intelligence is a fixed quantity, a quantity that cannot be increased. We must protest and react against this brutal pessimism: we must try to demonstrate that it is founded upon nothing.[1]

Goddard disagreed. He was convinced that the tasks were reliable indicators of intelligence, despite Binet's disclaimers. He and other researchers used the Binet-Simon scale as the basis of what is now known as IQ, or intelligence quotient. IQ is calculated by dividing a person's "mental age" as determined by the Binet-Simon scale by his or her chronological age and then multiplying by 100 to eliminate a decimal point. (A child with an IQ of 100 on such a test has a mental age equal to his or her chronological age.)

In the spring of 1913, Goddard decided to demonstrate the effectiveness of the Goddard-Binet test by sending two field workers to Ellis Island in New York harbor, the entry point for most immigrants. The two were told to "pass by the obviously normal" immigrant and choose individuals from the great mass of "average immigrants" for testing. They selected 35 Jews, 22 Hungarians, 50 Italians, and 45 Russians. Based on the results of those tests, Goddard claimed that 83 percent of the Jews, 80 percent of the Hungarians, 79 percent of the Italians, and 87 percent of the Russians were "feebleminded." In defense of these claims, he stated:

> Doubtless the thought in every reader's mind is the same as in ours, that it is impossible that half of such a group of immigrants could be feebleminded, but we know it is never wise to discard a scientific result because of its apparent absurdity. Not only are these figures representative of these ethnic groups as a whole, they are probably too small.[2]

When Goddard published his findings in 1917, a number of social workers and educators questioned his findings—particularly those that contradicted their own experiences. In a journal for social workers and others involved in "philanthropic charity work," Helen Winkler and Elinor Sachs wrote:

> "As stated," says Dr. Goddard's report in the Journal of Delinquency, "the physicians had picked out the obviously feebleminded, and to balance this we passed by the obviously normal." It would therefore seem that the group left was somewhat subnormal. But the paper goes on to say, "That left us the great mass of 'average immigrants.'" I always thought "average" meant normal, so that Dr. Goddard's group would from the start be below the par. This, and the fact that 148 persons altogether, or from twenty to fifty persons of each of the four nationalities represented, is entirely too small a number to constitute a fair sample upon which to base general conclusions, would make the results of the tests invalid if taken to have the significance the Survey clothes them with.
>
> But although Dr. Goddard slips up on his conclusions, he does not set out to prove the percentage of feeblemindedness among immigrants. The problems set for the experiment were: First, whether persons trained in work with the feebleminded could recognize, by simple inspection, the feebleminded immigrant; second, to what extent, if any, could mental tests successfully be applied to the detection of defective immigrants. . . .
>
> In his summary, the writer says, "It seems evident that mental

tests can be successfully used on immigrants, although much study is still necessary before a satisfactory scale can be developed." Following on the heels of this modest statement comes the assertion, "One can hardly escape the conviction that the intelligence of the average 'third class' immigrant is low, perhaps of moron grade."

The Department of Immigrant Aid of the Council of Jewish Women has been in daily contact with immigrants, particularly Jewish, and particularly women, girls, and children, who have much less opportunity for mental education than men and sometimes none at all. This daily contact does not bear out the statement of Dr. Goddard. In fact, the Department's statistics for the last fifteen months would show a contrary condition. Out of 2,549 Jewish women, girls, and children admitted during that time, only three were certified feebleminded.

. . . The conclusion of the Council of Jewish Women, drawn from its experience, is that out of the great bulk we have welcomed to our shores, the number of mental subnormals is inappreciable. . . .

In fact it is by no means agreed among psychologists that the Binet-Simon scale makes an accurate test of mental capacity, even though the examination may take into account the emotional state of the individual. In considering the value of the Binet test as applied to immigrants, we must take into account the fact that the test was originally designed for American children for the purpose of differentiating them into grades, and not to test capacity for mental development of peoples from different kinds of environment, with different languages, different education or lack of education.[3]

Even before Goddard published his findings, a number of his colleagues were also expressing their concerns about the test. J. E. Wallace Wallin, a clinical psychologist, gave two versions of the test to people he had known all of his life—individuals whose character and ability he could vouch for. At the 1915 meeting of the American Psychological Association, he gave a paper describing the results. It was later published in the 1916 *Journal of Criminal Law and Criminology* as "Who Is Feeble-Minded?" According to two versions of Goddard's tests, all of the "successful and wealthy" individuals Wallin tested in his hometown were "morons and dangerous feebleminded imbeciles." Wallin described one of those individuals in greater detail.

Mr. A, 65 years old, faculties well preserved, attended school only about 3 years in the aggregate; successively a successful farmer and business man, now partly retired on a competency of $30,000 (after considerable financial reverses from a fire), for ten years

> president of the board of education in a town of 700, superintendent or assistant superintendent of a Sunday school for 30 years; bank director; raised and educated a family of 9 children, all normal; one of these is engaged in scientific research (Ph.D.); one is assistant professor in a state agricultural school; one is assistant professor in a medical school (now completing thesis for Sc.D.); one is a former music teacher and organist, a graduate of a musical conservatory, but now an invalid; one a graduate of the [teacher training] department of a college; one is a graduate nurse; two are engaged in a large retail business; one is holding a clerical position; all are high school graduates and all except one have been one-time students in colleges and universities.
>
> . . . This man, measured by the automatic standards now in common use, would be hopelessly feeble-minded (an imbecile by the intelligence quotient), and should have been committed to an institution for the feeble-minded long ago. But is there anyone who has the temerity, in spite of the Binet "proof," to maintain, in view of this man's personal, social and commercial record, and the record of his family, that he has been a social and mental misfit, and an undesirable citizen, and should, therefore, have been restrained from propagation because of mental deficiency (his wife is still less intelligent than he)?[4]

Wallin urged his colleagues to join him in "completely rejecting the concept of the high grade moron as determined by the Binet scale from the standpoint of its moral and legal implications." A story in the Chicago newspapers provided Wallin with unexpected support. The papers revealed that Mary Campbell, a researcher in Chicago, had given the Goddard-Binet test to the mayor, his aides, and his opponents in the last election. Almost all of them were ranked as "morons." The American Psychological Association quickly resolved to discourage "the use of mental tests for practical psychological diagnosis by individuals psychologically unqualified for this work."

CONNECTIONS

Write a working definition of *intelligence.* Explain what the word means to you. Then add the meanings described in this reading. Record these definitions in your journal and add to them as you continue reading.

Alfred Binet wrote that "a French peasant may be normal in a rural community but feebleminded in Paris." Is the reverse equally true? Might a person who is normal in Paris be feebleminded in a rural community? What is Binet suggesting about the difficulties in defining the term *feebleminded*? The word *intelligence*? In measuring either? What "moral and legal implications" are implicit in the Binet scale?

What is an intelligence test? How is it different from an achievement test? An aptitude test? In small groups, write an example of a test question for each type of test. Share your questions with the class. Which questions were the easiest to write? To answer? To evaluate? Which were the hardest to create? To answer?

What questions do Winkler and Sachs raise about Goddard's methods? The authors describe their work with immigrants at Ellis Island. How do these experiences strengthen their arguments?

What does the word *normal* mean? *Average*? What assumptions is Goddard making when he directs his field workers to "pass by the obviously normal" immigrant and choose individuals from the great mass of "average immigrants" for testing? What assumptions do Winkler and Sachs make when they question his methods? What assumptions was Wallin making when he questioned the validity of the tests? What assumptions are reflected in his decision to test individuals whose history he knew rather than immigrants?

Winkler and Sachs use logic and personal experiences to challenge Goddard's conclusions. Goddard claimed his experiment at Ellis Island was "scientific." How does the very use of the word lend legitimacy and authority to a very unscientific survey? How does Goddard seem to define the word *scientific*? How do Winkler and Sachs define the term? How does Wallin seem to define it? What does it mean to you? To what extent does its use affect the way you regard a statement?

---

1. Quoted in *The Mismeasure of Man* by Stephen Jay Gould. W.W. Norton, 1981, pp.153-154.
2. "Mental Tests and the Immigrant" by H. H. Goddard. *Journal of Delinquency*, 1917, p. 266.
3. Letter to the Editor, *The Survey*, November 10, 1917.
4. "Who Is Feeble-Minded?" by J. E. Wallace Wallin. *Journal of Criminal Law and Criminology*, January, 1916.

Revising the Test

Reading 4

Even as Wallace Wallin and others were questioning the validity of the Goddard-Binet test, Lewis Terman, a professor of education at Stanford University, was creating a new version that would be later known as the Stanford-Binet test. It offered eugenicists a more reliable, less costly, and more efficient way of measuring the mental abilities of large groups of people.

To avoid Henry Goddard's errors, Terman normed every question—that is, he determined whether an "average" person could answer it by testing it on about 1000 children between the ages of 5 and 14 and 400 adults in his own community. Terman had difficulty finding enough adults to survey. In the end, he decided to treat anyone over the age of 14 as an adult. His 400 "adults" included 150 "tramps," 30 businessmen, 159 adolescent delinquents, and 50 high school students. Because the teenagers and the grown men got about the same number of items right on his test, Terman decided that "native intelligence, in so far as it can be measured by tests now available, appears to improve but little after the age of fifteen or sixteen years."

All of the individuals Terman tested were native-born Protestant Americans of Northern European descent. He made no secret of the fact that he eliminated "tests of foreign born children" "in the treatment of results." Commenting on the scores of immigrant children, Terman wrote:

> The tests have told the truth. These boys are ineducable beyond the merest rudiments of training. No amount of school instruction will ever make them intelligent voters or capable citizens. . . . They represent the level of intelligence, which is very, very common among Spanish-Indian and Mexican families of the Southwest and also among Negroes. Their dullness seems to be racial, or at least inherent in the family stocks from which they come.[1]

At first Terman's test, like the Goddard-Binet test, had to be administered individually by a trained examiner. An important breakthrough came in the spring of 1917, soon after the United States entered World War I. With the help of Henry Goddard and psychologist Robert Yerkes, Terman quickly devised a new version of the Stanford-Binet test—one that an untrained examiner could administer to hundreds of individuals at the same time. They planned to use the new test to determine which of the thousands of men recently drafted into the army were candidates for officer training and which were unfit to serve at all. Between May and June of 1917, the testers created eight Alpha and seven Beta tests. (Researchers often use the Greek letters alpha and beta to differentiate between two versions of the same test.) The Alpha tests were for draftees who could read English and the Beta for those who were illiterate or had little or no knowledge of English. While army officials were never completely convinced of the value of these tests, Terman, Goddard, and Yerkes had no doubts about their importance. They drew on the results of the so-called "army tests" again and again in their research. Yerkes wrote:

> Most of us are wholly convinced that the future of mankind depends in no small measure upon the development of the various biological and social sciences. . . . We must . . . strive increasingly for the improvement of our methods of mental measurement, for there is no longer ground for doubt concerning the practical as well as the theoretical importance of studies of human behavior. We must learn to measure skillfully every form and aspect of behavior which has psychological and sociological significance.[2]

Test 8

Notice the sample sentence: People *hear* with the *eyes* *ears* *nose* *mouth*

The correct word is *ears*, because it makes the truest sentence.

In each of the sentences below you have four choices for the last word. Only one of them is correct. In each sentence draw a line under the one of these four words which makes the truest sentence. If you can not be sure, guess. The two samples are already marked as they should be.

SAMPLES { People *hear* with the *eyes* *ears* *nose* *mouth*
{ *France* is in *Europe* *Asia* *Africa* *Australia*

1. The *apple* grows on a *shrub* *vine* *bush* *tree* — 1
2. *Five hundred* is played with *rackets* *pins* *cards* *dice* — 2
3. The *Percheron* is a kind of *goat* *horse* *cow* *sheep* — 3
4. The most prominent industry of *Gloucester* is *fishing* *packing* *brewing* *automobiles* — 4
5. *Sapphires* are usually *blue* *red* *green* *yellow* — 5
6. The *Rhode Island Red* is a kind of *horse* *granite* *cattle* *fowl* — 6
7. *Christie Mathewson* is famous as a *writer* *artist* *baseball player* *comedian* — 7
8. *Revolvers are made by* *Swift & Co.* *Smith & Wesson* *W. L. Douglas* *B. T. Babbitt* — 8
9. *Carrie Nation* is known as a *singer* *temperance agitator* *suffragist* *nurse* — 9
10. *"There's a reason"* is an "ad" for a *drink* *revolver* *flour* *cleanser* — 10
11. *Artichoke* is a kind of *hay* *corn* *vegetable* *fodder* — 11
12. *Chard* is a *fish* *lizard* *vegetable* *snake* — 12
13. *Cornell University* is at *Ithaca* *Cambridge* *Annapolis* *New Haven* — 13
14. *Buenos Aires* is a city of *Spain* *Brazil* *Portugal* *Argentina* — 14
15. *Ivory* is obtained from *elephants* *mines* *oysters* *reefs* — 15
16. *Alfred Noyes* is famous as a *painter* *poet* *musician* *sculptor* — 16
17. The *armadillo* is a kind of *ornamental shrub* *animal* *musical instrument* *dagger* — 17
18. The *tendon of Achilles* is in the *heel* *head* *shoulder* *abdomen* — 18
19. *Crisco* is a *patent medicine* *disinfectant* *tooth-paste* *food product* — 19

Partial example of a Alpha test.

Note: The test pictured below is an example of a Beta test. Each picture has a part missing. Identify the missing part in as many pictures as possible within three minutes. (The answers appear on page 180.)

Fig. 63.—Group Examination Beta, Form 0, Test 6, Picture Completion.

Example of a Beta test taken by army draftees in 1917.

CONNECTIONS

Take the test on the previous page by filling in the missing part of each drawing or identifying it on a separate sheet of paper. There is only one right answer for each test item. (Answers are provided at the end of the chapter.) Check your answers and then compare your score with those of your classmates.

A portion of the Alpha test is shown on page 157. To what extent is it like the Beta test? What differences seem most striking? How do both tests create the impression of scientific objectivity?

In Chapter 1, Martha Minow is quoted as saying, "When we simplify and sort, we focus on some traits rather than others, and we assign consequences to the presence and absence of the traits we make significant." What were the consequences of the way Americans defined intelligence in the early 1900s? What are the consequences today? How does this test seem to define intelligence? That is, what do you need to know to answer questions correctly? How do you define intelligence? How would you design a test to measure intelligence based on your definition?

1. *The Measurement of Intelligence* by Lewis Terman. Houghton Mifflin, 1916, p. 91.
2. Quoted in *The Mismeasure of Man* by Stephen Jay Gould. W. W. Norton, 1981, p. 193.

Fears of Declining Intelligence

Reading 5

Lewis Terman, Henry Goddard, and Robert Yerkes believed that American intelligence was declining. They saw the army tests as an opportunity to prove their theory. After testing over 1,750,000 army recruits, they and other experts took a sample of 160,000 for further analysis. In *A Study of American Intelligence*, Carl Brigham summarized what they learned from that analysis. Published in 1923, the book had a profound effect on popular attitudes toward immigrants and African Americans. Brigham, an assistant professor of psychology at Princeton University at the time and later president of the American Psychological Association, concluded:

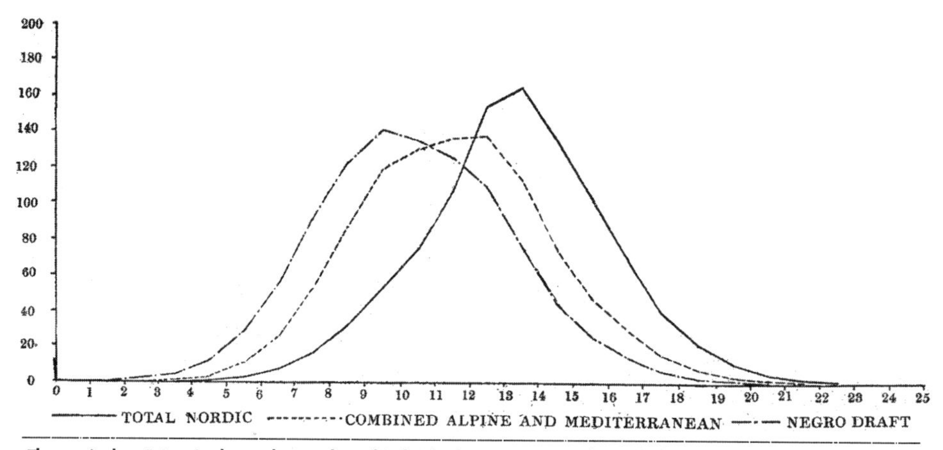

The vertical or "y" axis shows the number of individuals in 100,000s who took the army tests. The horizontal or "x" axis indicates the number of items they answered correctly.

> According to all evidence available, then, American intelligence is declining, and will proceed with an accelerating rate, as the racial admixture becomes more and more extensive. The decline of American intelligence will be more rapid than the decline of the intelligence of European national groups, owing to the presence here of the Negro. These are the plain, if somewhat ugly, facts that our study shows. The deterioration of American intelligence is not inevitable, however, if public action can be aroused to prevent it. There is no reason why legal steps should not be taken which would insure a continuously progressive upward evolution.
>
> The steps that should be taken to preserve or increase our present intellectual capacity must of course be dictated by science and not by political expediency. Immigration should not only be restrictive

> but highly selective. And the revision of the immigration and naturalization laws will only afford a slight relief from our present difficulty. The really important steps are those looking toward the prevention of the continued propagation of defective strains in the present population. If all immigration were stopped now, the decline of the American intelligence would still be inevitable. This is the problem which must be met, and our manner of meeting it will determine the future course of our national life.[1]

In the 1920s, most Americans—including many educators, religious leaders, politicians, and scientists—accepted Brigham's conclusions without question. After all, they confirmed what most of them already believed—some "races" were superior to others. Over the years, however, a number of researchers have challenged his conclusions. They point out:

1. From a sample of 160,000 army recruits, Brigham generalized to entire nations and ethnic groups.
2. Sample sizes varied among test groups. There might be 300 men in one group and 750 in another.
3. Testing conditions varied from one army camp to another. So did the instructions given to recruits. There were many zero scores, probably because soldiers did not understand the instructions.
4. There were discrepancies within and among groups. For example, African American recruits who lived in large cities in northern states tended to score higher than their southern, rural counterparts. Brigham said this resulted from a "better stock of Negro blood" in the North, but a simpler explanation might lie in the fact that African Americans in the North generally had more educational and economic opportunities than blacks in the South. (On the other hand, when Brigham found differences in scores between English-speaking and non-English-speaking Nordics, he attributed those differences to environment.)
5. Brigham offered no scientific definitions for how he determined who belonged to the Nordic, Alpine, Mediterranean, and Negro races. He simply adopted the conventional racist stereotypes that were used at the time.

CONNECTIONS

What were Brigham's conclusions? Why did many Americans accept them without question? How did Brigham's conclusions reinforce prejudices?

Copies of *A Study of American Intelligence* are still available in many libraries. It contains reproductions of several versions of the Alpha and Beta tests. Compare those tests to the ones in the previous reading. What questions does your research raise about the way Brigham used evidence? The conclusions he drew from that evidence? What experiences would you like to share with Brigham? What would you like him to know? What questions would you ask him?

Compare and contrast the way Brigham and Samuel Morton responded to discrepancies in their research (page 52). What similarities do you notice? How do you account for differences? What obstacles do scientists face in studying human beings? How might those obstacles be overcome? Why is it often so difficult to do so?

In response to those who argued that intelligence was declining and the nation needed more "geniuses," essayist H. L. Mencken wrote:

> The eugenicists constantly make the false assumption that a healthy degree of progress demands a large supply of first rate men. Here they succumb to the modern craze for mass production. Because a hundred policemen, or garbage men, or bootleggers are manifestly better than one, they conclude absurdly that a hundred Beethovens would be better than one. But this is not true. The actual value of genius often lies in its singularity.[2]

What does Mencken mean when he writes that value of genius often "lies in its singularity"? He also points out that composer Ludwig von Beethoven had a physical disability (deafness) and was "the grandson of a cook and the son of a drunkard." What is he suggesting about the relationship between genius and heredity?

In 1987, researcher James R. Flynn conducted a study of changes in IQ test scores over a 60-year period in such nations as Australia, Austria, Belgium, Brazil, Britain, Canada, China, Denmark, East Germany, France, Israel, Japan, Netherlands, New Zealand, Northern Ireland, Norway, Sweden, Switzerland, the United States, and West Germany. He discovered that IQ scores have not declined but increased in every country he surveyed, including the United States. No one knows why scores have gone up, but the changes have taken place too quickly to be attributed to evolution.[3] What questions does the so-called "Flynn effect" raise about the conclusions Brigham drew from the army tests? How does the "Flynn effect" challenge the idea that intelligence is simply a matter of heredity? Find out more about the "Flynn effect." What other eugenic assumptions does it challenge?

Invite a math teacher to explain to your class how to calculate frequency distributions. How do the mean, the median, and the mode differ from one another?

Use the illustration on page 160 to explain why a distribution of scores that clusters around the middle of a sample group is known as a "bell curve." What is the significance of such a distribution?

---

1. *A Study of American Intelligence* by Carl Brigham. Princeton University Press, 1923, p. 210.
2. Quoted in *Human Biodiversity* by Jonathan Marks. Aldine de Gruyter, 1995, pp. 91–92.
3. "Massive IQ Gains in 114 Nations: What IQ Tests Really Measure" by James R. Flynn. *Psychological Bulletin*, vol. 101 (1987), pp. 171-191.

Reading 6

For many people, Carl Brigham's *A Study of American Intelligence* confirmed their prejudices and therefore justified discrimination. By 1924, writes psychologist Fred Kuhlman, intelligence tests now had "an extra-scientific interest." "They decide the fate of thousands of human beings every year and are intimately related to social welfare in general." The tests were increasingly used to identify, segregate, and shame not only individuals but also entire groups of people. For example, Henry Fairfield Osborn, a trustee of Columbia University and president of the American Museum of Natural History, summarized the conclusions he and others drew from those data:

> We have learned once and for all that the Negro is not like us. So in regard to many races and subraces in Europe we learned some which we had believed possessed of an order of intelligence perhaps superior to ours were far inferior.[1]

In an article in the *Atlantic Monthly*, another writer noted that 89 percent of African Americans had tested as "morons." She made no mention of the fact that so did the mayor of Chicago and much of his staff (Reading 3). Instead she concluded that the "education of the whites and colored in separate schools may have justification other than that created by race prejudice."[2] Journalist Walter Lippmann challenged those who reached such conclusions:

> Without offering any data on all that occurs between conception and the age of kindergarten, they announce on the basis of what they have got out of a few thousand questionnaires that they are measuring the hereditary mental endowment of human beings. Obviously this is not a conclusion obtained by research. It is a conclusion planted by the will to believe. It is, I think, for the most part unconsciously planted.[3]

Sociologist W. E. B. DuBois, the first African American to earn a Ph.D., was also outraged by those who claimed that the tests "proved" that blacks were inferior:

> For a century or more it has been the dream of those who do not believe Negroes are human that their wish should find some scientific basis. For years they depended on the weight of the human brain, trusting that the alleged underweight of less than a thousand Negro brains, measured without reference to age, stature, nutrition,

or cause of death, would convince the world that black men simply could not be educated. Today scientists acknowledge that there is no warrant for such a conclusion and that in any case the absolute weight of the brain is no criterion of racial ability.

Measurements of the bony skeleton followed and great hopes of the scientific demonstration of racial inferiority were held for a while. But they had to be surrendered when Zulus and Englishmen were found in the same . . . class.

Then came psychology: the children of the public schools were studied and it was discovered that some colored children ranked lower than white children. This gave wide satisfaction even though it was pointed out the average included most of both races and that considering the educational opportunities and social environment of the races, the differences were measurements simply of the ignorance and poverty of the black child's surroundings.

Today, however, all is settled. "A workably accurate scientific classification of brain power" has been discovered and by none other than our astute army officers. The tests were in two sets for literates and illiterates and were simplicity itself. For instance among other things the literates were asked in three minutes "to look at each row of numbers below and on the next two dotted lines write the two numbers that should come next."

3	4	5	6	7	8	____	____
8	7	6	5	4	3	____	____
10	15	20	25	30	35	____	____
81	27	9	3	1	1/3	____	____
1	4	9	16	25	36	____	____
16	17	15	18	14	19	____	____
3	6	8	16	18	36	____	____

Illiterates were asked, for example, to complete pictures where the net was missing in a tennis court or a ball in a bowling alley!

For these tests were chosen 4730 Negroes from Louisiana and Mississippi and 28,052 white recruits from Illinois. The result? Do you need to ask? M. R. Trabue, Director, Bureau of Educational Service, Columbia University, assures us that the intelligence of the average southern Negro is equal to that of a 9-year-old white boy and that we should arrange our educational program to make "waiters, porters, scavengers, and the like" of most Negroes!

Is it conceivable that a great university should employ a man whose "science" consists of such utter rot?[4]

CONNECTIONS

How are Lippmann's remarks similar to Frederick Douglass's description of Samuel Morton and other "race scientists" as reasoning "from prejudice rather than from facts"? (page 52) What do both men suggest about the difference between "good" and "bad" science?

Douglass went on to say: "It is the province of prejudice to blind; and scientific writers, not less than others, write to please, as well as to instruct, and even unconsciously to themselves, (sometimes,) sacrifice what is true to what is popular. Fashion is not confined to dress; but extends to philosophy as well—and it is fashionable now, in our land, to exaggerate the differences between the Negro and the European." To what extent do Lippmann's comments suggest that scientific writers in the early 1900s continued to "sacrifice what is true to what is popular"? How do those writers perpetuate myths and misinformation about "race"?

Based on the test questions cited by DuBois, how do you think the testers defined intelligence? How does DuBois seem to define it? To what extent does the Beta test shown in Reading 4 support DuBois's conclusions about the inadequacies of the test?

---

1. Quoted in *The Mismeasure of Man* by Stephen Jay Gould. W. W. Norton, 1981, p. 231.
2. Ibid., p. 231.
3. Ibid., p. 174.
4. "Race Intelligence" by W. E. B. DuBois. *Crisis*, July 1920, pp. 1181–1183. ©1920 *Crisis*.

Reading 7

By the 1920s, intelligence tests were increasingly used to determine who was "worthy" of a variety of educational and employment opportunities. As early as 1922, educator John Dewey warned against any test "which under the title of science" reduces "the individual to a numerical class; judges him with reference to capacity to fit into a limited number of vocations ranked according to present business standards; assigns him to a predestined niche and thereby does whatever education can do to perpetuate the present order." His warning was largely ignored. Companies continued to use them to determine which applicants to hire. Colleges and universities like Oberlin, the University of Illinois, Brown, Purdue, and Southern Methodist in Dallas depended on them to screen incoming freshmen. So did a number of high schools.

As mentioned earlier, between 1880 and 1920, school enrollment in the United States increased by more than 600 percent, from about 200,000 students in 1880 to over 1.5 million by 1920. Much of the increase was a direct result of state laws that required children to attend school until at least the age of 14.

As the number of students increased seven-fold, school officials struggled to educate youngsters with diverse abilities from a wide variety of backgrounds. Many schools used test scores to assign students to particular classes. In practice, this meant keeping immigrant and African American students from courses that might prepare them for higher education and educating them only for unskilled, low-paying jobs. For example, Ellwood Cubberly, a professor of education at Stanford and an eugenicist, wrote in 1916:

> Our schools are factories in which the raw products are to be shaped and fashioned into products. . . . The specifications for manufacturing come from the demands of 20th century civilization, and it is the business of the school to build its pupils according to the specifications laid down. This demands good tools, specialized machinery, and continuous measurement of production.[1]

Although not every one accepted these ideas, they shaped the way thousands of communities across the nation allocated funds for education. They also reinforced old myths about race and ethnicity that fed prejudice, limited opportunity, and undermined self-esteem. Poet Paisley Rekdal writes:

> At sixteen, my mother loads up red tubs of noodles, teacups

Illustration from the *American School Board Journal*, 1922.

chipped and white-gray as teeth, rice clumps that glue themselves to the plastic tub sides or dissolve and turn papery in the weak tea sloshing around the bottom. She's at Diamond Chan's restaurant, where most of her cousins work after school and during summer vacations, some of her friends, too. There's Suzy at the cash register, totaling up bills and giving back change, a little dish of toothpicks beside her and a basket of mints that taste like powdered cream. A couple of my mother's cousins are washing dishes behind the swinging kitchen door, and some woman called Auntie #2 (at her age, everyone is Auntie and each must take a number) takes orders at a table of women that look like Po Po's mah-jongg club. They don't play anymore. They go to the racetrack.

The interior of Diamond Chan's restaurant is red: red napkins, red walls, red carp in the tank and in signature seals on the cheap wall hangings. Luck or no luck, it's like the inside of an esophagus. My mother's nails are cracked, kept short by clipping or gnawing, glisten only when varnished with the grease of someone else's leftovers. Still she keeps working here, it is repetitive action, the chores that keep her from thinking. The money my mother earns will soon get sucked into the price of a pink cashmere sweater for Po Po's birthday, along with a graduation photo of herself, also in a pink sweater, pearls, her face airbrushed fog-rose at the cheeks and mouth.

Graduation? Unlike her brothers, she knows she's going to college. Smith, to be exact, though without the approval of the school counselor. "Smith is . . . expensive," the counselor told my mother only yesterday, which is why my mother is slightly irritated now, clomping around under the weight of full tubs of used dishes. "Smith is not for girls like you." What does she plan to be when she grows up? "A doctor?" my mother suggests. Um, no. "Nursing. Or teaching, perhaps, which is even more practical. Don't you think?"

My mother, who is practical above all things, agreed.

So it's the University of Washington in two years with a degree in education. Fine. She slams down full vials of soy sauce onto each table. . . . Smith is not for girls like her. . . .

It is not, my mother would argue, that she could be denied the dream of Smith so much that someone should tell her that she could be denied it. My mother knows the counselor was hinting at some limitation my mother would prefer to ignore. Still, she is whiter than white, should intelligence be considered a pale attribute. Deep down she understands she has a special capacity for work; she likes it, she's good at it, she excels at school and its predictable problems. Here is a discipline entirely lacking in the spirits of whatever loh fan may sneer or wonder at her in study hall; to be told by a fat, dyed-blond guidance counselor she may be inferior? The monkey calling the man animal.[2]

Malcolm Little was also a top student in his high school in Lansing, Michigan. He kept his grades high even though he too held a part-time job in a restaurant. He worked as a dishwasher. In his autobiography, Little recalls a conversation with one of his teachers.

Somehow, I happened to be alone in the classroom with Mr. Ostrowski, my English teacher. He was a tall, rather reddish white man and he had a thick mustache. I had gotten some of my best marks under him, and he had always made me feel that he liked me. . . .

He told me, "Malcolm, you ought to be thinking about a career. Have you been giving it thought?"

The truth is, I hadn't. I never have figured out why I told him, "Well, yes sir, I've been thinking I'd like to be a lawyer." Lansing certainly had no Negro lawyers—or doctors either—in those days, to hold up an image I might have aspired to. All I really knew for certain was that a lawyer didn't wash dishes, as I was doing.

Mr. Ostrowski looked surprised, I remember, and leaned back in his chair and clasped his hands behind his head. He kind of half-smiled and said, "Malcolm, one of life's first needs is for us to be realistic. Don't misunderstand me now. We all here like you, you know that. But you've got to be realistic about being a nigger. A lawyer—that's no realistic goal for a nigger. You need to think about something you can be. You're good with your hands—making things. Everybody admires your carpentry in shop work. Why don't you plan on carpentry? People like you as a person—you'd get all kinds of work."

The more I thought afterwards about what he said, the more uneasy it made me. It just kept treading around in my mind.

What made it really begin to disturb me was Mr. Ostrowski's advice to others in my class—all of them white. Most of them told him they were planning to become farmers. But those who wanted to strike out on their own, to try something new, he had encouraged. Some, mostly girls, wanted to be teachers. A few wanted other professions, such as one boy who wanted to become a county agent; another, a veterinarian; and one girl wanted to be a nurse. They all reported that Mr. Ostrowski had encouraged what they had wanted. Yet nearly none of them had earned marks equal to mine.

It was a surprising thing that I had never thought of it that way before, but I realized that whatever I wasn't, I was smarter than nearly all of those white kids. But apparently I was still not intelligent enough, in their eyes, to become whatever I wanted to be.

It was then that I began to change—inside. [3]

Malcolm Little is better known today as Malcolm X. In 1952, he changed his name when he converted to Islam.

CONNECTIONS

How does Rekdal express her mother's anger at the guidance counselor? How is it reflected in her statement that her mother is "whiter than white, should intelligence be considered a pale attribute"? In her description of counselor as "the monkey calling the man animal"?

In a famous study, an educational psychologist randomly selected a number of elementary-school students. The psychologist told the children's teachers that the tests showed these students were likely to do better in school the coming year than they had ever done before. By the end of the year, the students had indeed done better, as measured by the grades they received and by their teachers' comments. They did better, the psychologist concluded, because their teachers expected them to do better. Charles Davenport believed that teachers ought to have "a record of inherited capabilities or performances of close relatives" in order to better predict the abilities of each child. What does this study suggest about drawbacks of such a system? What does the study suggest about the relationship between what others expect of us and what we become?

After his encounter with Ostrowski, Malcolm X recalls, "It was then that I began to change—inside." What kinds of changes might such an incident inspire? How did a similar incident seem to shape the identity of Paisley Rekdal's mother? How do you like to think you would react to such an incident?

After interviewing writer Maya Angelou for a television series on creativity, journalist Bill Moyers reflected on the importance of having people in our lives who have faith in us, even when we lack faith in ourselves. Angelou told him of a trauma that left her silent and described how she eventually regained her voice thanks to her grandmother's love and the compassion of a neighbor. In assessing what these two women meant to the child, Moyers states:

> For the inner life to flourish everyone needs to be touched by someone. . . . With Maya Angelou, it was a grandmother who loved her vastly and a radiant black angel who read Dickens to a little girl not quite turned eight. They signified her worth, they said, "You matter," they turned her suffering rage upward and brought the poet to life. It is not a scientifically certifiable fact that with each child born into the world comes the potential to create. It is rather a statement of faith. But I can't imagine any declaration more important for our society to make.[4]

What do the accounts written by Paisley Rekdal and Malcolm X suggest happens when a society fails to make such a statement of faith?

---

1. *Public School Administration* by Ellwood Cubberly. Houghton Mifflin, 1916, p. 338.
2. *The Night My Mother Met Bruce Lee* by Paisley Rekdal. Copyright © 2000 by Paisley Rekdal. Used by permission of Pantheon Books, a division of Random House, Inc.
3. *The Autobiography of Malcolm X* as told to Alex Haley. Ballantine, 1965, pp. 35–37.
4. Interview with Maya Angelou. *Creativity* with Bill Moyers. Public Affairs Television.

Reading 8

In the early 1900s, scholars like Carl Brigham routinely used racist stereotypes in their work. As a result, their research bolstered old myths and misinformation by offering "scientific proof" that intelligence is related to morality; that some races are superior to others; and that African Americans "are intellectually inferior to whites and can only be educated within clear limits." The few who dared to ask questions often had difficulty getting heard. African Americans had a particularly difficult time. It was no accident that W. E. B. DuBois's criticism of the Army tests was published only in *Crisis*, the journal of the National Association for the Advancement of Colored People (NAACP).

In 1924, Horace Mann Bond, the director of education at Langston University, reviewed Carl Brigham's *A Study of American Intelligence*. His review was also published in *Crisis*. He wrote:

> The manner in which these tests and their results are being regarded should cause serious concern on the part of the Negro Intellectual, for in many cases they have ceased to be scientific attempts to gain accurate information and have degenerated into funds for propaganda and encouragement for prejudice. It should therefore be the aim of every Negro student to be in possession of every detail of the operation, use and origin of these tests, in order that he might better equip himself as an active agent against the insidious propaganda which like its prototypes, seeks to demonstrate that the Negro is intellectually and physically incapable of assuming the dignities, rights and duties which devolve upon him as a member of modern society. . . .
>
> Why should Negroes from Northern states possess larger increments of intelligence than Negroes from Southern states? Mr. Brigham says that this is because the more intelligent have immigrated northward; a very pretty explanation, but not one which can be taken to justify the fact. There is only one obvious explanation; the Negro from the North, because of infinitely superior home, civil, and above all school conditions, has been favored by environment in just as great a degree as his Southern brother has been deprived of the same. . . .
>
> Thus with the list of other "inferiorities" so confidently affirmed by Mr. Brigham and others of his school. Invariably a perusal of those nationalities whom he classes as inferior will be found to have

> a close correlation existing between the sums of money expended for education and their relatively low standing. . . .
>
> Only recently an investigator working from the University of Texas proclaimed the fact that he found the Negro children possessed but 75 per cent of the average intelligence native to whites. Further investigation revealed the following facts: In that special locality, the whites, with a school population of 10,000, were expending on an average $87 per capita for the education of their children. The Negro children received a per capita of $16; and yet this Texas psychologist believes he has unearthed a brief for Negro inferiority. . . .
>
> No, it is not with Intelligence Tests that we have any quarrel; in many ways they do represent a fundamental advance in the methodology of the century. It is solely with certain methods of interpreting the results of these tests that we, as scientific investigators, must differ. So long as intelligence tests are administrated, correlated, and tabulated solely with the subjective urge subdued, and with a certain degree of common sense as to their interpretation, we can never criticize them.
>
> But so long as any group of men attempts to use these tests as funds of information for the approximation of crude and inaccurate generalizations, so long must we continue to cry "Hold!" To compare the crowded millions of New York's East Side with the children of some professional family on Morningside Heights indeed involves a great contradiction; and to claim that the results of the tests given to such diverse groups, drawn from such varying strata of the social complex, are in any wise accurate, is to expose a fatuous sense of unfairness and lack of appreciation of the great environmental factors of modern urban life.[1]

With no funding from charitable foundations and no support from the academic community, the studies were too small to alter the "conventional wisdom." They did, however, encourage other scholars. Otto Klineberg, a psychologist and a student of anthropologist Franz Boas (Chapter 3) was among the first to seek evidence in support of Bond's criticisms. In his *A Study of American Intelligence*, Carl Brigham had used the results of the Army tests to argue that "Nordics" [Northern Europeans] are mentally superior to "Mediterranean" and "Alpine" peoples. To test that claim, Klineberg administered performance tests in ten of the "purest" Nordic, Mediterranean, and Alpine villages he could find in Europe. The three groups showed no significant differences in the kinds of abilities the test measured.

Next Klineberg turned his attention to differences between the scores of black and white Americans. In Europe, he noticed that people who lived in cities did better on the tests than those who lived in rural areas. In the United States, blacks who lived in large cities scored higher on the average than both black and white groups in rural communities in the South. Brigham and other eugenicists explained the phenomenon by arguing that people in urban areas scored higher because the more intelligent people tended to leave rural areas for the city.

To test that idea, Klineberg examined school records of black children in three southern cities to determine whether those who went north were brighter than those who stayed behind. He also gave IQ tests to southern-born African Americans who had lived in New York City for various lengths of time to see if the environment made a difference. In 1935, Klineberg wrote:

> The superiority of the northern over the southern Negroes to approximate the scores of the whites, are due to factors in the environment, and not to selective migration. The school records of those who migrated did not demonstrate any superiority over those who remained behind. The intelligence tests showed no superiority of recent arrivals in the North over those of the same age and sex who were still in southern cities. There is, on the other hand, very definite evidence that an improved environment, whether it be the southern city as contrasted with the neighboring rural districts, or the northern city as contrasted with the South as a whole, raises the test scores considerably; this rise in "intelligence" is roughly proportionate to length of residence in the more favorable environment.[2]

As Klineberg and others challenged racist assumptions in one study after another, a number of psychologists and other social scientists began to doubt their findings. In 1928, Henry Goddard admitted that a person who has an IQ in the 70s was probably not a "moron." He also acknowledged that many people who did poorly on his tests were able to learn, grow, and even improve their scores. In time, he even backed away from his claim that the "feebleminded" were a grave threat to the general public.

In 1930, Carl Brigham also had second thoughts about his work. In a public retraction of the conclusions he reached in *A Study of American Intelligence*, he stated, "Comparative studies of various national and racial groups may not be made with existing tests. . . . One of the most pretentious of these comparative racial studies—the writer's own—was without foundation."[3]

CONNECTIONS

What does an intelligence test measure? What do the results reveal? Why does Bond call Brigham's interpretation of the IQ tests "propaganda"? Why did he believe it was important for African American students to know the details of the "operation, use, and origin" of these tests? How important was it that other students also know those details?

How did Klineberg's research challenge the "conventional wisdom" about racial differences? About the relationship between heredity and intelligence? About the relationship between environment and intelligence?

What kind of evidence does Bond use to challenge Brigham's conclusions? What evidence does Klineberg use? From your own experience, what kinds of proofs are most likely to alter perspectives? Inspire a reassessment of a long-held belief?

What questions does this reading raise about the importance of dissent? How did dissident voices make their views known then? How do they make their views known today? Is it enough to just speak out?

In the late 1800s, a group of German anthropologists tried to determine whether there were racial differences between Jewish and "Aryan" children. After studying nearly seven million students, the society concluded that that the two groups were more alike than different. According to historian George Mosse, the survey should have ended racist thinking in Europe. Instead, he concludes, "The idea of race had been infused with myths, stereotypes, and subjectivities long ago, and a scientific survey could change little. The idea of pure, superior races and the concept of a racial enemy solved too many pressing problems to be easily discarded."[4] What do Mosse's comments suggest about the difficulty of overcoming myths about race? How do his comments support the view that what people believe is true is often more important than the truth?

---

1. "Intelligence Tests and Propaganda" by Horace Mann Bond. Copyright © *Crisis*, June, 1924 (vol. 28, #23).
2. Quoted in *In the Name of Eugenics* by Daniel Kevles. Harvard University Press, 1995, p. 138.
3. "Intelligence Tests of Immigrant Groups" by Carl Brigham, *Psychological Review* 37, 1930, p. 165.
4. *Toward the Final Solution: A History of European Racism* by George Mosse. Fertig, 1978, p. 92.

Intelligence Testing Today

Reading 9

Journalist Walter Lippmann was one of the earliest critics of intelligence testing. He was not sure whether the test measured "the capacity to pass tests or the capacity to deal with life, which we call intelligence." Regardless, he wrote, the examiner "is testing the complex result of a long and unknown history, and the assumption that his questions and his puzzles can in fifty minutes isolate abstract intelligence is, therefore, vanity."

In the 1920s, many researchers dismissed Lippmann's criticism as uninformed because he was not a psychologist. They also ignored scholars like Horace Mann Bond and Otto Klineberg. Today scholars are not as quick to discount such critiques. Wendy M. Williams, an associate professor of human development, explains why.

> With no formal schooling to speak of, [my grandfather] could build anything, from a dollhouse to a real house, from scratch, without plans. He also could fix anything—kitchen appliances, cars, children's toys, radios, televisions, you name it. He even published a book of his poems when he was in his 70s. He was not clever, however, at taking I.Q. tests, which he confronted in grade school, in the military, and when he looked for a job when he was in his early 20s. He hated taking the tests; he was made anxious by the clock ticking as he worked, and he found it confusing and unnatural to think in terms of abstractions, be they mathematical, pictorial, or verbal.
>
> Because of his performance on tests, my grandfather did not consider himself very intelligent. Neither did the teachers, military recruiters, and job-placement personnel who used the test scores: They reduced my grandfather's intelligence to a simple, relatively low number on a page and labeled him "slow." The I.Q. tests that my grandfather took in the 1930s—versions of which are still in use today—were created to determine which children failing in school were doing so because of low intelligence, and which were failing for other reasons. Through questions about the meaning of words or paragraphs, mathematical problems, visual patterns, and so forth, these tests measured intelligence in terms of the number of problems a person could solve, compared with the average for other people of the same age.
>
> Throughout our society, we still use I.Q. tests, and their close surrogates such as the SAT [the Scholastic Aptitude Test], in the belief

that they provide a meaningful measure of a person's innate intelligence and capacity for success in intellectual tasks. We all know the considerable weight these tests are given throughout education, as well as in hiring and promotion decisions in the workplace. But scholars still have not explained how, if I.Q. tests tell us the most important things we need to know about a person's intelligence, we can account for my grandfather and the many others like him, who are competent and successful in so many domains in the real world.

This is the issue that my colleagues and I have studied in our attempt to democratize the concept of intelligence, by including in it more and different types of abilities and talents. While we have been conducting our research, other scholars working in the same area have demonstrated that I.Q. tests' reputation as an ultimate seal of approval was premature.

For example, consider the work of James Flynn, a political scientist at the University of Otago, in New Zealand. He proved that I.Q. scores have risen sharply over the past 60 or more years in all 20 nations for which data exist. In fact, a person born in 1877 whose score put him or her in what was then the 90th percentile on a widely used reasoning test would, with exactly the same number of correct answers, rank in only the 5th percentile of people born in 1967. (Flynn proved this by examining the raw numbers of correct answers on the same tests used over time. Most researchers rely on "normed scores," which are adjusted to keep the average score on a test constant from year to year, and which thus cannot accurately be compared over time.) . . .

We learned two things from Flynn's work: First, a high I.Q. score does not necessarily mean intelligence, nor does a low score mean stupidity. Second, whatever the test measures is highly mutable. Flynn is fond of saying that, if we take I.Q. scores seriously as meaningful predictors of intelligence, our grandparents would have been unable to understand the rules of baseball. Given the rapidity of the changes Flynn reported, genetics could not be responsible, and so researchers have focused on aspects of culture, as well as on health and nutrition, in attempting to explain why people today are markedly outscoring their ancestors.

One possible cultural factor is that people are increasingly familiar with the material on certain types of I.Q. tests. My grandfather's generation rarely encountered anything in their everyday lives even remotely resembling the items on such tests. Today, however, mazes, puzzles, and other games that are thinly disguised versions of items from actual I.Q. tests appear on cereal boxes and on place mats at fast-food restaurants. People play with toys such as Rubik's Cube. Some computer screen-saver programs are strikingly similar to other kinds of intelligence tests: The complex patterns dancing around the screen closely resemble the . . . most popular test of reasoning ability. Is it any wonder that today's kids outperform my grandfather's generation?

But the more important question is: Does this greater exposure to material similar to that on the tests make today's children and adults smarter in any meaningful way than earlier generations of test takers? I think not. The intellectual accomplishments of people in past eras are awe-inspiring, and the challenges and hardships that they had to overcome were extraordinary. Looking back on these accomplishments should make us cautious in interpreting the significance of I.Q. scores as predictors of likely success in the real world.

Perhaps the reason that so many individuals with low or moderate I.Q.'s, such as my grandfather, are so successful in their daily lives can be found in recent research that has broadened the concept of intelligence. Researchers today are demonstrating empirically the importance of many abilities that are not measured on I.Q. tests. Consider studies that my colleagues and I have conducted to assess practical and creative thinking in business, the military, and elementary and middle schools.

We wanted to know why some business managers with M.B.A.'s from prestigious graduate schools alienate their subordinates virtually overnight, why some military leaders lose the respect of their soldiers and subordinate officers by adhering to formal doctrines even in situations where they are not adequate, and why some bright children hand in boring compositions after the deadline and then react with surprise when they receive low grades. We found that all of these people lack practical intelligence—an ability essential to success that differs from the more "academic" intelligence measured by I.Q. tests, and which is largely independent of it.

We learned that practical intelligence consists of three types of abilities—managing oneself, managing others, and managing the organization or environment in which one works, such as a school, corporation, or hospital. Each ability is important in a unique way, and each contributes to real-world success. People may be strong in one type of practical intelligence and weak in another, although, generally, being savvy about managing organizations builds on the abilities to manage oneself and others. Importantly, traditional measures of

I.Q. tell us little about who has and does not have the three types of practical intelligence.

Where are scientists headed in our search to understand intelligence? Increasingly, we think in terms of types and facets of intelligence that lead to success in specific contexts: social intelligence, emotional intelligence, creative intelligence, and practical intelligence. We look at people's ability to manage their lives by motivating and organizing themselves to perform effectively. We consider people's ability to get along with their employees, peers, supervisors, and teachers. Often, it is those types of intelligence, as much as I.Q. scores, that determine success or failure in education and in the workplace, especially among people with a similar range of I.Q. scores.

Historically, a person's intelligence was reduced to a single number. Today, that number still holds sway in many admissions offices, but the realization is growing that we need to characterize and measure more of the abilities that are important to adult success. We owe the next generation a broader and more relevant battery of tests, designed to measure the many varied abilities that contribute to success in the real world. Better tests will lead to the admission of applicants with a wider variety of skills, thus diversifying further the pool of talent available to our society.

As we look ahead to the demographic changes under way and recognize the need to distribute educational and employment opportunities fairly and broadly, it becomes even more essential for us to assess people's capabilities accurately. We need a conception of intelligence that encompasses my grandfather's talents. The most successful leaders in business, the professions, and other enterprises know how to define workable goals and motivate themselves to accomplish them; they know how to "read" and motivate other people; and they know how to distinguish solutions that work in the real world from ones that work only in books—all abilities that current I.Q. tests do not measure.

This is not to say that success on an I.Q. test does not provide meaningful information; it is just that other types of success matter, too. It should not escape us that the technological developments on which our society depends may require types of intelligence—practical and creative, for example—that are different from those emphasized in our standardized tests. The science of understanding intelligence thus may progress farther and faster by recognizing the wisdom of our grandparents.[1]

CONNECTIONS

If IQ tests tell us the most important things about a person's intelligence, asks Williams, how do we "account for my grandfather and the many others like him"? How does she answer her own question? How do you think Goddard or Brigham would answer it? How would you answer it?

What does the work of James Flynn add to our understanding of intelligence? How does it undermine the belief that intelligence is a matter of genes?

Scientist Jacob Bronowski writes that "every judgment in science stands on the edge of error, and is personal." How does Williams's account illustrate that idea? What is the difference between viewing one's work as "on the edge of error" as opposed to "on the edge of truth"?

Chapter 1 raised the question of what do we do with a difference. How does Williams seem to answer that question? How does her answer differ from the way eugenicists answered that question? What do your answers suggest about what it means to "democratize" intelligence?

In 1999, a Princeton molecular biologist inserted in mice a gene that codes for a protein in brain cells associated with memory. Because the experimental animals performed better than the control mice on tests of learning, the media claimed that the researcher had located "the smart gene" or the "IQ gene." How did the reporters seem to define intelligence? How important is that definition?

Write a definition of *intelligence* based on the working definition you created as you read this chapter. Research recent efforts to define the term and use your findings to revise or expand your definition. You may want to focus on the work of such scholars as James Flynn, Daniel Goleman, who stresses the importance of emotional intelligence, or Howard Gardner, who writes of multiple intelligences. How does their work deepen your understanding of intelligence? What new questions does their research raise?

---

1. "Democratizing I.Q." by Wendy M. Williams. Copyright ©1998 by *The Chronicle of Higher Education*, May 5, 1998. Permission granted by author.

Answers to Beta test, page 158.

1. Mouth
2. Eye
3. Nose
4. Spoon
5. Chimney
6. Ear
7. Filament
8. Stamp
9. Strings
10. River
11. Trigger
12. Tail
13. Leg
14. Shadow
15. Ball (in hand)
16. Net
17. Forearm
18. Horn
19. Arm (in mirror)
20. Diamond

6. Toward Civic Biology

The prime duty of the good citizen of the right type is to leave his or her blood behind in the world, and we have no business permitting the perpetuation of the wrong type.

Theodore Roosevelt

Chapter 5 revealed how eugenicists tried to provide not only a scientific rationale for long-standing prejudices but also someone to blame for all of society's ills. Chapter 6 explores how eugenicists and their supporters translated their beliefs about difference into public policy in the early 1900s. Using the language of public health and progressive reform, they argued that progress required "cleansing" the nation of citizens of the "wrong type." As Charles R. Van Hise, then president of the University of Wisconsin, explained:

> We know enough about agriculture so that the agricultural production of the country could be doubled if the knowledge were applied. We know enough about disease so that if the knowledge were utilized, infectious and contagious diseases would be substantially destroyed in the United States within a score of years; we know enough about eugenics so that if the knowledge were applied, the defective class would disappear within a generation.[1]

Many Americans liked the idea of a seemingly scientific approach to the nation's social problems. Few questioned whether a democratic government has a right to take such action. This chapter explores two "eugenic laws" that gave government a say in the most fundamental choices a person makes—the selection of a mate and the decision to have children. Anti-miscegenation laws and statutes requiring forced sterilization were passed by elected legislatures and signed into law by elected governors. Many judges considered these laws in keeping with the Constitution.

How did eugenicists win support for laws that labeled friends, neighbors, even relatives as "defective" or "racially inferior"? How did they convince people that government has the right, even a duty, to interfere in the most personal decisions an individual can make? These questions are central to Chapter 6. The chapter also considers the significance of the way we as individuals and members of a society define our universe of obligation—the circle of individuals and groups to whom we feel obligated.

---

1. Quoted in *Eugenics: Hereditarian Attitudes in American Thought* by Mark H. Haller. Rutgers University Press, 1985, p. 76.

What Did You Learn in School Today?

Reading 1

By 1928, eugenics was part of the curriculum in most high schools as well as in 376 institutions of higher learning—including Harvard, Columbia, Cornell, Brown, Wisconsin, and Northwestern. According to one survey of 41 textbooks, nearly 90 percent of all high school biology textbooks published between 1914 and 1948 endorsed the movement. Students also learned that some among them were a threat to society in their psychology, sociology, anthropology, and home economics classes.[1]

What did high school students learn about eugenics and how did it shape their lives? In *The New Civic Biology*, a textbook first published in 1914, author George William Hunter alerted young people to the "menace of feeblemindedness" and the value of "breeding the best with the best" by using the language of science to heighten real fears about the spread of diseases and the threat of possible disabilities.

> Since our knowledge of heredity has been increased, the demand has become more urgent that we do something to prevent the race from handing down diseases and other defects, and that we apply to man some of the methods we employ in breeding plants and animals. This is not a new idea; the Greeks in Sparta had it, Sir Thomas More wrote of it in his Utopia, and today it has been brought to us in the science of eugenics. The word comes from the Greek word eugenes, which means well born. Eugenics is the science of being well born, or born well, healthy, fit in every way. A tendency to cancer, or tuberculosis, or chorea, or feeblemindedness, is a handicap which it is not merely unfair, but criminal, to hand down to posterity.
>
> **Two notorious families**
>
> Studies have been made on a number of different families in the country, in which mental and moral defects were present in one or both of the parents as far back as was possible to trace the family. The "Jukes" family is a notorious example. "Margaret, the mother of criminals," is the first mother in the family of whom we have record. Up to 1915 there were 2094 members of this family; 1600 were feebleminded or epileptic, 310 were paupers, more than 300 were immoral women, and 140 were criminals. The family has cost the state of New York more than $2,500,000, besides immensely

lowering the moral tone of the communities which the family contaminated.

Another careful investigation (up to 1912) concerned the "Kallikak" family. This family was traced to the union of Martin Kallikak, a young solider of the War of the Revolution, with a feebleminded girl. She had a feebleminded son, who had 480 descendants. Of these 33 were sexually immoral, 24 confirmed drunkards, 3 epileptics, and 143 feebleminded. The man who started this terrible line of immorality and feeblemindedness later married a normal Quaker girl. From this couple a line of 496 descendants was traced, with no cases of feeblemindedness. The evidence and the moral speak for themselves!

Parasitism and its Cost to Society

Hundreds of bad families such as those described exist today, spreading disease, immorality, and crime to all parts of this country. The cost to society of such families is very severe. Just as certain animals or plants become parasitic on other plants or animals, these families have become parasitic on society. They not only do harm to others by corrupting, by stealing, and by spreading disease, but they are actually protected and cared for by the state out of public money. It is estimated that between 25% and 50% of all prisoners in penal institutions are feebleminded. They take from society, but they give nothing in return. They are true parasites. . . .

Blood Tells

Eugenics shows us, on the other hand, in a study of families in which brilliant men and women are found, that the descendants have received the good inheritance from their ancestors. . . . Although we do not know the precise method of inheritance, we do know that musical and literary ability, calculating ability, remarkable memory, and many other mental and physical characters are inheritable and "run in families." The Wedgewood family, from which three generations of Darwins have descended, and the Galton family are examples of scientific inheritance; the Arnolds, Hallams, and Lowells were prominent in literature; the Balfours were political leaders; the Bach and Mendelssohn families were examples showing inheritance of musical genius. A comparison of fathers' and sons' college records at Oxford University shows [that] . . . fathers who did well had sons who did well also. It is said that 26 out of 46 men chosen to the Hall of Fame of New York University had distinguished relatives. Blood does tell!

How to Use Our Knowledge of Heredity

> Two applications of this knowledge of heredity stand out for us as high school students. One is in the choice of a mate, the other in the choice of a vocation. As to the first, no better advice can be given than the old adage, "Look before you leap." If this advice were followed, there would be fewer unhappy marriages and divorces. Remember that marriage should mean love, respect, and companionship for life. The heredity of a husband or wife counts for much in making this possible. And, even though you are in high school, it is only fair to yourselves that you should remember the responsibility that marriage brings. You should be parents. Will you choose to have children well born? Or will you send them into the world with an inheritance that will handicap them for life?[2]

The implications of Hunter's questions were reinforced in popular newspapers, magazines, and books. For many Americans, they had a very real meaning. In 1939, a minister in Pontiac, Michigan, sent the following letter to the Eugenics Record Office at the suggestion of a physician in Chicago:

> A few years ago two young men came to our city from a state several hundred miles distance. These brothers lived in our home and shared our devotional and church life and we have learned to love them as one of our family.
>
> Then one day, unsolicited, word came to me that these boys have a strain of black blood in their veins. This seemed impossible to me since there were no Negro characteristics apparent to me. They have no thick lips. Their hair is light brown and their eyes light blue and their complexion is fair. But I carefully sought facts and when I was fairly certain from these, I approached the boys and they too informed me that it was so and that they learned the truth when they were in high school. I believe they have told me the truth and they are facing the problem, for which they are not responsible, in a very courageous manner.
>
> These boys, 24 and 28 years of age, have met girls and I discover that they are contemplating marriage. The girls know the conditions and the girls' parents also know it. The young people and the mother of one of the girls have come to me for advice. There is no feeling of animosity between the parents and the young men. There is a deep sympathetic desire to do what is right. Their questions are: Is there any assurance that children would not revert back to black? Can they be certain of birth control methods? Is sterilization the only

positive and right procedure? If the man submits to such an operation could the sex act still be practiced with satisfying results? Do you have any scientific data on hand to give us a helpful report?

These young men both contemplate sterilization and have asked me to investigate for them. I plan to do that today, but I shall wait for word from you before any final step is taken.

I am enclosing a statement of the family history as accurately given as the boys know from the father's side. The facts are not fully known on the mother's side. If there is a strain on the mother's side the boys think that it is probably with the American Indian.

If there are any questions and statements of fact that you desire and I can be of any help, be assured that I shall gladly do anything possible to get to the truth in this matter.

It seems unfortunate that a public announcement of the engagement and date of anticipated wedding has been made. Probably some postponements will have to be made so if we could have some word soon it would be deeply appreciated.

It appears now that they will go on with their plans and our ultimate question is, should sterilization be done?[3]

CONNECTIONS

The author of *The New Civic Biology* uses such words as *we*, *us*, and *our* throughout the passage. Who are *we*? Who is one of *us*? Members of a society are part of a universe of obligation—a community made up of the individuals and groups toward whom members have obligations, toward whom they believe the rules of society apply, and whose injuries call for amends. Whom does the author of *The New Civic Biology* seem to exclude? How does he want readers to define their "universe of obligation"? What does the letter suggest may be the consequences of such a definition?

The New Civic Biology is a textbook. How are textbooks like other informational books? What distinguishes them from other non-fiction? How does that distinction affect the content of the book? The way it is written? Compare the passage from *The New Civic Biology* with a modern biology textbook. What similarities do you notice? What differences seem striking?

Propaganda refers to words or images that are designed to spread a message or promote a cause. What is the message of the passage quoted from *The New Civic Biology*? At whom is that message aimed? To what emotions does it appeal?

What is the moral or message of the passage? How does the author's use of such words as *criminal*, *notorious*, *parasite*, and *immorality* underscore that message? What terms are clearly defined in this account? What terms are only vaguely defined? To what extent is the passage an example of propaganda?

The New Civic Biology states that "a tendency to cancer, or tuberculosis, or chorea, or feeblemindedness, is a handicap which it is not merely unfair, but criminal, to hand down to posterity." What is the difference between an "unfair" handicap and one that is "criminal"? How do you think students who had such "tendencies" in their families may have reacted to such statements? What is the author suggesting about their future?

According to Hunter, whose achievements are signs of good heredity—those of the men of the family or the women? How do you think views like his may have affected the way students saw themselves, their families, and their classmates?

How may textbooks like Hunter's have shaped the way the brothers responded to the discovery that they were at least partly African American? Why do you think the minister turned to a physician in Chicago for advice rather than the Bible, another minister, or even a local doctor? Why do you think the physician referred him to the Eugenics Record Office?

What can we learn about the two brothers from the letter? Why do you think that aspects of their identity were kept secret from them until they were in high school? How do you explain their willingness to undergo sterilization rather than pass on their genes to another generation? What does their willingness to do so suggest about the power of eugenics? The danger it posed to ordinary people?

There is no record of a response to the minister's letter. How do you think eugenicists like Harry Laughlin would respond to the young men's "desire to do what is right"? How would you respond? What questions would you like to ask the two young men? What would you like to tell them and their fiancées? The minister who wrote the letter?

What values and beliefs shape the letter? What does it suggest about the power of myths and misinformation to shape a person's life? Racism is often thought of as hatred towards a minority group. Yet the minister expresses no hatred. Indeed he cares deeply about the two young men he describes. Is he a racist?

---

1. *Inheriting Shame: The Story of Eugenics and Racism in America* by Steve Seldon. Teacher's College Press, 1999, p. 64.
2. *The New Civic Biology* by George William Hunter. American Book Company, 1914, pp. 398–401.
3. Courtesy of Harry H. Laughlin Papers, Pickler Memorial Library, Truman State University, Kirksville, Missouri.

Reading 2

In challenging students to choose a mate carefully, the author of *The New Civic Biology* (Reading 1) implied that it was an individual choice. And for some individuals like the young men from Michigan described in the reading, it was. In other parts of the United States, the government had a voice in that decision, as Richard Loving and Mildred Jeter would discover.

Loving and Jeter grew up in Virginia's rural Caroline County in the 1950s. They met at a dance and dated for a few years before deciding to marry. After a wedding in Washington, D.C., they returned to Virginia to start a family. Historians Peter Irons and Stephanie Guitton write:

> Six weeks later, the Lovings had a terrible shock. Sheriff Garnet Brooks arrived with a warrant directing him to bring "the body of said Richard Loving" before a judge. He dragged the Lovings out of bed. And what was their crime? Rich was white and Mildred had mixed black and Indian ancestry. Their marriage violated a Virginia law providing that "if any white person intermarry with a colored person"—or vice versa—each party "shall be guilty of a felony" and face prison terms of five years.
>
> The Lovings pleaded guilty to avoid prison. Judge Leon Bazile suspended a one-year sentence if they agreed to leave Virginia for twenty-five years. The Lovings moved to Washington, but they were country people and couldn't adjust to city life. They came back to Caroline County and lived a fugitive life for nine years, sheltered by family and friends and raising three small kids. "I never expected . . . such a beating," Rich said later. "It was right rough."
>
> Rich appealed for help to Attorney General Robert Kennedy in 1963. Kennedy sent his letter to the American Civil Liberties Union, which recruited two Virginia lawyers, Philip Hirschkop and Bernard Cohen. They [argued] that the Lovings' conviction [violated] the Fourteenth Amendment's guarantee of "equal protection of the laws" to Americans of all races. Civil rights and church groups [supported] the appeal.[1]

In 1967, the case now known as *Loving v. Virginia* reached the Supreme Court. Ten days after the couple's ninth wedding anniversary, the justices issued a unanimous opinion: Virginia's law was unconstitutional. This ruling also overturned anti-miscegenation laws—laws that banned marriages between whites and individuals of other "races"—in fifteen other states. Chief Justice Earl Warren stated the court's opinion:

> Marriage is one of the "basic civil rights of man," fundamental to our very existence and survival. . . . To deny this fundamental freedom on so unsupportable a basis as the racial classifications embodied in these statutes . . . is surely to deprive all the State's citizens of liberty without due process of law. The Fourteenth Amendment requires that the freedom of choice to marry may not be restricted by invidious racial discriminations. Under our Constitution, the freedom to marry, or not marry, a person of another race resides with the individual and cannot be infringed by the state.

Anti-miscegenation laws date back to colonial times. The first such statute was passed by the Maryland General Assembly in 1691. Other colonies followed suit. These laws were an American invention.There was no ban on interracial marriage in England at the time. By the late 1800s, 38 states had anti-miscegenation statutes. As late as 1924 these laws were on the books in 29 states. Anti-miscegenation laws varied greatly in the way they defined whom one could and could not marry. In a legal brief filed in *Loving v. Virginia*, the National Association for the Advancement of Colored People (the NAACP) commented on the inconsistencies in these laws:

> In Mississippi, Mongolian-White marriages are illegal and void, while in North Carolina they are permitted. . . . In Arkansas, a Negro is defined as any person who has in his or her veins "any Negro blood whatever"; in Florida, one ceases to be a Negro when he has less than "one-eighth of African or Negro blood," and in Oklahoma, anyone not of the "African descent" is miraculously transmuted into a member of the white race.

A number of states updated their anti-miscegenation laws in the 1920s. The Virginia Racial Integrity Act of 1924, which the Lovings violated, is a good example. Its sponsors used eugenic arguments to justify restrictions. They argued that interracial relationships are "dysgenic unions" in which "the superior group (whites) risks polluting their germ plasm with inferior hereditary traits." Lothrop Stoddard, a lawyer and self-proclaimed eugenics expert, supported the proposed law. He told Virginia lawmakers:

> White race purity is the cornerstone of our civilization. Its mongrelization with non-white blood, particularly with Negro blood, would spell the downfall of our civilization. This is a matter of both

national and racial life and death, and no efforts would be spared to guard against the greatest of all perils—the perils of miscegenation.[2]

On March 20, 1924, state lawmakers passed the Virginia Racial Integrity Act by a wide margin, and the governor signed it into law. The Virginia law remained on the books until 1967 when the Supreme Court overturned it in *Loving v. Virginia*. The law stated in part:

> **Section 1-14 of the Virginia Code:**
> Colored persons and Indians defined—Every person in whom there is ascertainable any Negro blood shall be deemed and taken to be a colored person, and every person not a colored person having one fourth or more of American Indian blood shall be deemed an American Indian. . . .
>
> **Section 20-54 of the Virginia Code:**
> Intermarriage prohibited; meaning of term 'white persons.'—It shall hereafter be unlawful for any white person in this State to marry any save a white person, or a person with no other admixture of blood than white and American Indian. For the purpose of this chapter, the term 'white person' shall apply only to such person as has no trace whatever of any blood other than Caucasian; but persons who have one-sixteenth or less of the blood of the American Indian and have no other non-Caucasic blood shall be deemed to be white persons. . . .
>
> **Section 20-58 of the Virginia Code:**
> Leaving State to evade law —If any white person and colored person shall go out of this State, for the purpose of being married, and with the intention of returning, and be married out of it, and afterwards return to and reside in it, cohabiting as man and wife, they shall be punished as provided in §20-59, and the marriage shall be governed by the same law as if it had been solemnized in this State. The fact of their cohabitation here as man and wife shall be evidence of their marriage.
>
> **Section 20-59 of the Virginia Code:**
> Punishment for marriage.—If any white person intermarry with a colored person, or any colored person intermarry with a white person, he shall be guilty of a felony and shall be punished by confinement in the penitentiary for not less than one nor more than five years.

Walter Plecker, a physician and the director of the Virginia Board of Vital Statistics, was responsible for the enforcement of the law in the early 1900s. He and his staff relied on birth certificates, marriage licenses, tax records, and gossip to decide who was white and who was not. Plecker "corrected" birth certificates if he thought a person was trying to "pass" as white. He targeted Native Americans in the belief that they were really blacks trying to pass as something else. The pride Plecker took in his work is evident in a letter he wrote in 1943, during World War II: "Our own indexed birth and marriage records, showing race, reach back to 1853. Such a study has probably never been made before. . . . Hitler's genealogical study of the Jews is not more complete."[3]

CONNECTIONS

The word *miscegenation* comes from two Latin words—*miscere*, which means "mix" and *genus* for "race." What is miscegenation? What is the purpose of an anti-miscegenation law?

Over the years, the Supreme Court has identified a number of rights that are "so rooted in the traditions and conscience of our people as to be ranked as fundamental." Why does the Court consider the right to marry "fundamental"? What other rights are viewed as "fundamental"?

Why do anti-miscegenation laws regard interracial marriages as a "public health" issue? What arguments did Lothrop Stoddard offer the General Assembly in favor of the proposed Virginia Racial Integrity Act? What scientific arguments might you offer to counter his argument? What moral or philosophical arguments were offered in support of the law? How would you respond to them? If possible, refer to particular provisions in the law.

American Indian groups in Virginia, including the Monacan tribe in Amherst County, are still feeling the consequences of Plecker's interpretation of the Virginia Racial Integrity Act. They have been unable to persuade the federal government to recognize them as tribes because Plecker erased all evidence of their heritage. Some have called it "a paper genocide." What does it mean to have your heritage—a part of your identity—erased?

---

2. *Richmond Times-Dispatch*, Feb. 13, 1924, p. 1.

3. Letter from Walter Plecker to John Collier, Office of Indian Affairs, April 6, 1943.

Reading 3

In the early 1900s, anti-miscegenation laws were not the only marriage laws enacted or amended according to "eugenic principles." Many states also outlawed or restricted marriages in which one or both partners were "feebleminded," "insane," epileptic, or had a venereal disease.

In 1895, Connecticut became the first state to outlaw marriages that involved "defective" persons. The new law called for imprisonment for up to three years for a party to a marriage or an extra-marital relationship in which one partner was "feebleminded," an epileptic, or an "imbecile." The only exception was for women over the age of 45—that is, women who were past their childbearing years. Over the next 20 years, 24 states enacted similar laws. By the mid-1940s, 41 states had laws prohibiting the marriage of the mentally ill and the "feebleminded." Seventeen states banned marriages to epileptics and alcoholics.

Eugenicists applauded these measures but believed that they addressed only a part of the problem. Therefore they urged that the states pass laws that identified and then sterilized women and men "unfit" to reproduce. In 1913, a writer for *The Psychological Bulletin* argued, "The burden of supporting these people must not rest any more heavily upon the normal race." Since the "unfit" were becoming too numerous to be segregated, he insisted, "the only thing to do is to sterilize them. With procreation stopped, the matter would be practically under control in a generation."[1]

Eugenicists were not the first to favor laws that would make it impossible for the "unfit" to have children. In the United States, the practice began with prison officials who, in the belief that "criminalism" is inherited, saw sterilization as a deterrent to crime. At the turn of the 20th century, a prison official in Indiana carried out dozens of vasectomies without the legal authority to do so. He later reported, "It occurred to me that this would be a good method of preventing procreation in the defective and physically unfit."[2] Eugenicists actively encouraged state lawmakers to make such sterilizations legal. In 1907, Indiana became the first state to permit involuntary sterilization if a committee of experts decided that a prisoner should not be allowed to have children. Other states followed Indiana's lead in sterilizing criminals as well as the disabled.

By 1924, 21 states had laws permitting involuntary sterilization. The impetus to pass these laws came from eugenicists like Harry Laughlin, the superintendent of the Eugenics Record Office. In 1914, Laughlin wrote a Model Sterilization Law that was circulated widely in the United States and Europe. He also testified personally before state legislatures that were considering sterilization laws or arranged for other eugenicists to do so. Their scientific expertise and prestige had an impact on lawmakers.

Laughlin's Model Sterilization Law states in part:

> An Act to prevent the procreation of persons socially inadequate from defective inheritance, by authorizing and providing for the eugenical sterilization of certain potential parents carrying degenerate hereditary qualities.
>
> (a) A socially inadequate person is one who by his or her own effort, regardless of etiology or prognosis, fails chronically in comparison with normal persons, to maintain himself or herself as a useful member of the organized social life of the state; provided that the term socially inadequate shall not be applied to any person whose individual or social ineffectiveness is due to the normally expected exigencies of youth, old age, curable injuries, or temporary physical or mental illness, in case such ineffectiveness is adequately taken care of by the particular family in which it occurs.
>
> (b) The socially inadequate classes, regardless of etiology or prognosis, are the following (1) Feeble-minded: (2) Insane, (including the psychopathic): (3) Criminalistic (including the delinquent and wayward): (4) Epileptic: (5) Inebriate (including drug habitués): (6) Diseased (including the tuberculous, the syphilitic, the leprous, and others with chronic, infectious and legally segregable diseases): (7) Blind (including those with seriously impaired vision): (8) Deaf (including those with seriously impaired hearing): (9) Deformed (including the crippled): and (10) Dependent (including the orphans, ne'er-do-wells, the homeless, tramps and paupers). . . .
>
> (f) A potential parent of socially inadequate offspring is a person who, regardless of his or her own physical, physiological or psychological personality, and of the nature of the germ-plasm of such person's co-parent, is a potential parent at least one fourth of whose possible offspring, because of the certain inheritance from said parent of one or more inferior or degenerate physical, physiological or psychological qualities would, on the average, according to the demonstrated laws of heredity, most probably function as socially inadequate persons; or at least one-half of whose possible offspring would receive from said parent, and would carry in the germ-plasm but would not necessarily show in the personality, the genes or genes-complex for one or more inferior or degenerate physical, physiological or psychological qualities, the appearance of which

quality or qualities in the personality would cause the possessor thereof to function as a socially inadequate person under the normal environment of the state.

Section 3. Office of State Eugenicist
There is hereby established for the State of .... the office of State Eugenicist, the function of which shall be to protect the state against the procreation of persons socially inadequate from degenerate or defective physical, physiological or psychological inheritance.

Section 4. Qualifications of State Eugenicist
The State Eugenicist shall be a trained student of human heredity, and shall be skilled in the modern practice of securing and analyzing human pedigrees: and he shall be required to devote his entire time and attention to the duties of his office as herein contemplated. . . .

Section 7. Duties of State Eugenicist
It shall be the duty of the State Eugenicist: (a) To conduct field-surveys seeking first-hand data concerning the hereditary constitution of all persons in the State who are socially inadequate personally or who, although normal personally, carry degenerate or defective hereditary qualities of a socially inadequate nature, and to cooperate with, to hear the complaints of, and to seek information from individuals and public and private social-welfare, charitable and scientific organizations possessing special acquaintance with and knowledge of such persons, to the end that the State shall possess equally accurate data in reference to the personal and family histories of all persons existing in the State, who are potential parents of socially inadequate offspring, regardless of whether such potential parents be members of the population at large or inmates of custodial institutions, regardless also of the personality, sex, age, marital condition, race or possessions of such persons.[3]

From the start, sterilization laws were controversial. In seven states, local and state judges overturned these laws. A number of them argued that they violated the Fourteenth Amendment to the U.S. Constitution, which grants every citizen due process and equal protection under the law. Other judges noted that the laws unfairly singled out "feebleminded persons" in state institutions for sterilization, while leaving other individuals who were mentally defective alone. Still others believed that sterilization violated the Eighth Amendment to the Constitution, which bans "cruel and unusual punishments."

CONNECTIONS

How is a "socially inadequate person" defined in Section (a) of the "Model Sterilization Law"? What terms in this definition might prove difficult to define? Who does the law hold responsible for the care of socially inadequate persons?

No one has ever proved that there is a genetic link between "feeblemindedness" and poverty or crime. Even physical disabilities might be the result of a variety of factors. In 1910, psychiatrist Smith Ely Jellife warned:

> Is it logical to take such an enormous complex of conditions as all the psychoses and try to make them fit in one artificial box? It is the same way with epilepsies. . . . There is no one epilepsy. Convulsions could arise from a hard blow to the head, a motor area thrombus provoked by infection, or poisoning. . . . Is there any heredity here—or chance of it? If eugenics is to be correctly started, we must sharpen up our conceptions, and that very markedly.[4]

What categories in the model law does Jellife seem to challenge? What aspects of the law does he seem to accept without question? What other causes for "genetic conditions" does he suggest? What is his attitude toward the eugenics movement as a whole? (Ironically, Harry Laughlin could have been sterilized under the statute he drafted. He developed epilepsy as an adult.)

Laws requiring sterilization violated the basic rights of the victims. How did eugenicists and their supporters seem to justify those civil rights violations? What arguments might you offer in support of the victims?

Compare the assumptions in Harry Laughlin's "Model Sterilization Law" with the passage from Davenport's *Heredity in Relation to Eugenics* reprinted on pages 75-76. What similarities do you notice? What relationship between science and government do the two men seem to favor?

What are the duties of the state eugenicist, according to the model law? What families are likely to be investigated? What protections does the law provide for their privacy?

---

1. Quoted in *In Search of Human Nature* by Carl Degler. Oxford University Press, 1991, pp. 46–47.
2. Ibid.
3. *Eugenical Sterilization in the United States* by Harry Laughlin. Psychopathic Laboratory of the Municipal Court of Chicago, 1922, pp. 446-448.
4. Quoted in *In the Name of Eugenics* by Daniel Kevles. Harvard University Press, 1995, p. 49.

Reading 4

Critics of forced sterilization laws believed that they violated rights guaranteed in the U.S. Constitution. In 1924, eugenicists and their supporters decided to find out if the laws were constitutional. To do so, they needed someone who could challenge the law in the courts. They chose Carrie Buck of Virginia. At the age of 17 years old, she was pregnant and unmarried. Her mother, Emma, an inmate at the Lynchburg Colony for Epileptics and Feebleminded, was rumored to have been a prostitute. Carrie was classified as "feebleminded" and after her child was born, she was committed to the Lynchburg Colony. Officials were convinced that they now knew everything worth knowing about her.[1]

A simple check of state records would have revealed that Emma Buck and her husband were legally married at the time Carrie was born, although they separated when she was very young. Unable to support Carrie after she and her husband parted, Emma placed the four-year-old in foster care. The child was sent to live with a Mr. and Mrs. J. T. Dobbs. She did chores for the couple and attended school through the sixth grade. She kept up with her classmates and was promoted every year. According to school records, her sixth-grade teacher characterized Buck's work and behavior as "very good."

Like most poor children in rural Virginia in the first years of the twentieth century, Buck received a sixth-grade education. After leaving school, she continued to live with Dobbses and work in their home. She attended church and sang in the choir. In the early 1920s, a nephew of Mrs. Dobbs joined the household, possibly to help with farm work much as Buck helped with the housework. In the summer of 1923, when Buck was about 16, the nephew raped her while his aunt and uncle were away from home.

When Carrie Buck became pregnant, the Dobbses tried to commit her to the Lynchburg Colony by claiming that she had appeared "feebleminded" since the age of ten or eleven. Later they said she was "peculiar" since birth, even though she did not come to live with them until much later. State officials did not question these claims. After all, Carrie Buck fit their stereotype of a "feebleminded" girl. She was poor, pregnant, and uneducated.

On March 28, 1924, Carrie Buck gave birth to a daughter, whom she named Vivian. A few months later, Carrie was admitted to the Lynchburg Colony. Not long after her arrival, Virginia passed a law allowing involuntary sterilization of those labeled as "feebleminded." Officials at the Lynchburg Colony decided to sterilize Carrie Buck under the new law with the approval of Albert Priddy, the superintendent of the colony. But first, he and his colleagues arranged for her to appeal the decision in the Virginia courts. Although the appeal was in her name, Carrie Buck had no voice in the process. Priddy and other eugenicists were in charge. They hired an attorney for her as well as one for themselves. The two lawyers were in constant contact with one another and with Priddy before and during trial proceedings even though such collaborations are unethical.

The case, later known as *Buck v. Bell,* was first heard in the Circuit Court for Amherst County on November 18, 1924. At the trial Aubrey Strode, the lawyer for Priddy and the Lynchburg Colony, offered "scientific evidence" that Carrie Buck ought to be sterilized. The evidence came from the Eugenics Record Office and was prepared by Harry Laughlin. It stated:

> Carrie Buck: Mental defectiveness evidenced by failure of mental development, having chronological age of 18 years, with a mental age of 9 years, according to Stanford Revision of Binet-Simon Test: and of social and economic inadequacy; has record during life of immorality, prostitution, and untruthfulness: has never been self-sustaining; has had one illegitimate child, now about 6 months old and supposed to be a mental defective. . . .
>
> This girl comes from a shiftless, ignorant, and worthless class of people and it is impossible to get intelligent and satisfactory data, though I have had Miss Wilhelm, of the Red Cross of Charlottesville, try to work out her [family] line. . . .
>
> Further evidence of the hereditary nature of Carrie Buck's feeblemindedness and moral delinquency consists in the fact that at a very early age of four years she was taken from the bad environment furnished by her mother and given a better environment by her adopted mother. . . . The family history record and the individual histories, if true, demonstrate the hereditary nature of the feeblemindedness and moral delinquency in Carrie Buck. She is therefore a potential parent of socially inadequate or defective offspring.[2]

Laughlin's statement was based on information provided by the colony. He never met Buck. It was important to the colony's case to show that Buck was likely to pass on defective traits to her children. After watching her seven-month-old daughter for a short time, a nurse decided that the baby was "not quite normal." Based on this testimony, the judge decided that Carrie's mother, Carrie herself, and her infant daughter were all "socially inadequate."

Irving Whitehead, Buck's lawyer, did little on her behalf. He called no witnesses to dispute Laughlin or other "experts" who favored sterilization. Not surprisingly, a judge upheld the decision to sterilize Carrie Buck. Whitehead promptly

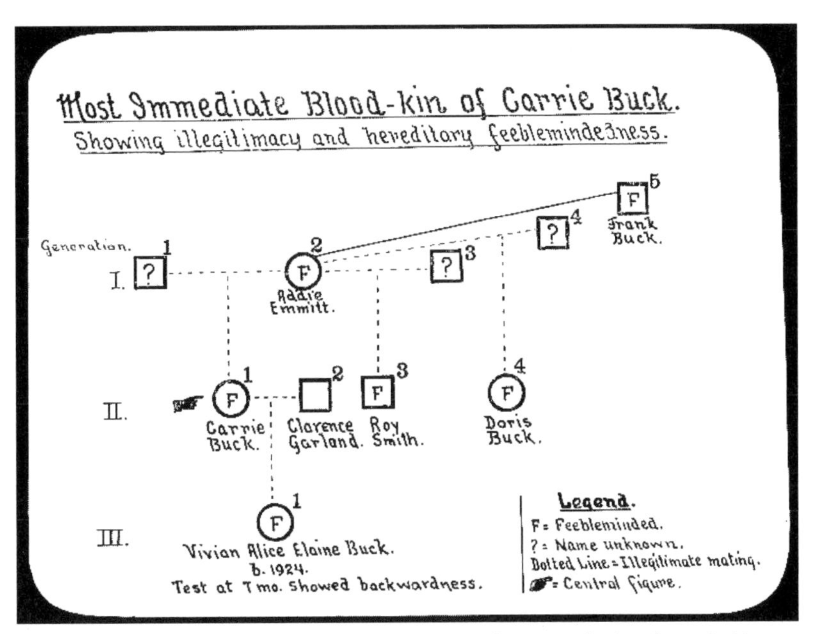

A pedigree chart that Harry Laughlin offered as evidence of Carrie Buck's hereditary "feeblemindedness."

filed an appeal on her behalf in the Virginia Court of Appeals. It was just eight pages long, compared with the 44-page document the colony's lawyers prepared. In November 1925, the appeals court also ruled against Buck.

In April of 1927, the case reached the U.S. Supreme Court. By then, Albert Priddy was dead. The new superintendent of the Lynchburg Colony was his former assistant, a Dr. Bell. So the case that began as *Buck v. Priddy* went to the Supreme Court as *Buck v. Bell.* The justices saw only the records from the original trial and the appeals court. Based solely on what they read in the court transcripts, they voted 8-1 to uphold the sterilization of Carrie Buck. Justice Oliver Wendell Holmes, Jr. stated:

> The case comes here upon the contention that the statute authorizing the judgment is void under the Fourteenth Amendment as denying to the plaintiff in error [Carrie Buck] due process of law and the equal protection of the laws.
>
> Carrie Buck is a feebleminded white woman who was committed to the State Colony above mentioned in due form. She is the daughter of a feebleminded mother in the same institution, and the mother of an illegitimate feebleminded child. She was eighteen years old at the time of the trial of her case in the Circuit Court in the latter part of 1924. . . . The commonwealth [of Virginia] is supporting in various institutions many defective persons who if now discharged

would become a menace but if incapable of procreating might be discharged with safety and become self-supporting with benefit to themselves and to society; . . . experience has shown that heredity plays an important part in the transmission of insanity, imbecility, etc. . . .

There can be no doubt that so far as procedure is concerned the rights of the patient are most carefully considered, and as every step in this case was taken in scrupulous compliance with the statute and after months of observation, there is no doubt that in that respect the plaintiff in error has had due process at law. . . .

We have seen more than once that the public welfare may call upon the best citizens for their lives. It would be strange if it could not call upon those who already sap the strength of the State for these lesser sacrifices, often not felt to be such by those concerned, in order to prevent our being swamped with incompetence. It is better for all the world, if instead of waiting to execute degenerate offspring for crime, or to let them starve for their imbecility, society can prevent those who are manifestly unfit from continuing their kind. The principle that sustains compulsory vaccination is broad enough to cover cutting the Fallopian tubes. . . . Three generations of imbeciles are enough.

The ruling had important consequences. Carrie Buck was sterilized in October 1927. She was paroled from the colony shortly after the operation with the stipulation that she report to officials annually. Over the years, Buck worked at odd jobs in households and on farms. She married, was widowed, and later remarried. She died in a nursing home in 1983. People who knew her remarked on her kindness and recalled her enjoyment of reading. Her daughter Vivian died from an infection in 1932, at the age of eight. School records show that she was a good student who made the honor roll at least once.

In 1928, Virginia officials also sterilized Carrie Buck's sister. She was told that the operation was to remove her appendix. Only in 1980 did she learn why she was never able to have a child. "I broke down and cried," she said. "My husband and me wanted children desperately. We were crazy about them. I never knew what they'd done to me."

The ruling encouraged other states to enact sterilization laws. By 1930, 24 states had passed similar measures and about 60,000 people were sterilized under these statutes. Virginia alone sterilized more than 7,500 people between the Supreme Court ruling in 1927 and 1972 when the law was finally replaced.

Most of the people who were sterilized came from poor or working-class backgrounds, much like Carrie Buck's. Patients from well-to-do families were cared for at home or private facilities. They rarely underwent sterilization. African Americans and other people of color were also unlikely to be sterilized—mainly because they were not admitted to public mental hospitals or institutions.

CONNECTIONS

How did Virginia's sterilization law view Carrie Buck and the other women in her family? To what extent did the law place them outside the state's "universe of obligation"?

Review the Supreme Court's decision and identify key elements in Holmes's justification for upholding the Virginia law. Who, in his view, are the "best citizens"? What is the implication of his use of such phrases as "a menace" and "swamped with incompetence"? To what extent did Buck receive "due process at law"? The Supreme Court relied on Laughlin and other eugenicists to make its decision. What scientific studies might have led to a very different decision? What does Holmes mean when he says that sterilization is a sacrifice "often not felt to be such by those concerned"? What assumptions is he making? What evidence provided in the reading might have altered his opinion? Historian Carole R. McCann writes of his decision:

> In effect, the Court gave the government the right to determine which women were competent to become mothers. Although Carrie Buck, the woman in the case, was poor and white, the Court's decision implicitly endorsed elitism and racism. It sanctioned the eugenic logic behind sterilization laws that defined fitness by class and race as much as by intelligence or character.[3]

What is elitism? Racism? In what sense does the decision endorse either or both? How does the decision define "fitness"?

What arguments might a more able and impartial lawyer have made on Carrie Buck's behalf? How might such a lawyer have challenged the scientific testimony in support of her sterilization?

The Supreme Court is mainly concerned with constitutional issues. What constitutional claim did Carrie Buck's lawyer make? The first section of the Fourteenth Amendment states:

> All persons born or naturalized in the United States, and subject to the jurisdiction thereof, are citizens of the United States and of the State wherein they reside. No state shall make or enforce any law which shall abridge the privileges or immunities of citizens of the United States; nor shall any State deprive any person of life, liberty, or property, without due process of law; nor deny to any person within its jurisdiction the equal protection of the laws.

How do you think the amendment applies to this case? Why does Holmes believe it does not apply?

The Lynchburg Story is a powerful documentary film on the Carrie Buck case and its legacy. The video adds an important element to this case by bringing it down from a legal and scientific plane into the real lives of the people involved.

---

1. This account is based upon an article by Paul A. Lombardo, "Three Generations, No Imbeciles: New Light on Buck v. Bell," *New York University Law Review*, vol. 60 (April, 1985), pp. 30–61.
2. The Supreme Court of Appeals of Virginia, No. 1700, *Carrie Buck vs Dr. J.H. Bell*, pp. 40–42.
3. "Eugenics" by Carole R. McCann in *The Reader's Companion to U.S. Women's History* ed. by Wilma Mankiller, Gwendolyn Mink, et. al. Houghton Mifflin, 1998, p. 179.

Reading 5

Forced sterilizations violated basic civil rights. In recent years a number of victims have demanded apologies and compensation for the damages done to them in the name of eugenics. Journalist Bill Baskervill describes one such victim:

> His state labeled him a "mental defective" and surgically sterilized him.
>
> His nation honored him as a war hero, awarding him the Bronze Star for valor, the Purple Heart and the Prisoner of War Medal for service in World War II.
>
> Now the Virginia House of Delegates has refused to apologize to Raymond W. Hudlow and the thousands of other Virginians, mostly teen-agers and young adults, who were sterilized under the state's eugenics program. Instead, the House . . . voted to express its "profound regret" for the General Assembly's action 77 years ago that led to forced sterilizations.
>
> "Does this man [Hudlow] deserve an apology, or just regrets?" said Phil Theisen, president of the Lynchburg Depressive Disorders Association. Theisen is a leading advocate for a state apology.
>
> Virginia officials, acting under a eugenics law that served as a model for the rest of the nation, tried to purify the white race from 1924 to 1979 by targeting virtually any human shortcoming they believed was a hereditary disease that could be stamped out by surgical sterilization. Such maladies included mental illness, mental retardation, epilepsy, criminal behavior, alcoholism and immorality.
>
> In 1941, Hudlow was a frightened16-year-old who became caught up in Virginia's eugenics frenzy that led to the forced sterilization of about 7,500 people.
>
> His crime: repeatedly running away from home to avoid beatings by his father.
>
> "Every time my father beat me I ran away. He beat me half to death," Hudlow said in a recent interview at his mobile home near Lynchburg.
>
> When his father told the "welfare lady" "he couldn't control me," Hudlow's reproductive fate was sealed.
>
> "I was picked up by the sheriff at home. He handcuffed me and took me" to the Virginia Colony for Epileptics and Feebleminded near Lynchburg, where most of Virginia's sterilizations were performed.

On June 17, 1942, Amherst County Circuit Judge Edward Meeks granted the colony's request to sterilize Hudlow, identified in the court order as an "inmate" of the colony.

Hudlow, now 75, remembers the day the colony eugenicists came for him.

"They just came and got me before I woke up one morning. They wheeled me and throwed me up on the operating table. They put straps around my waist and chest, spread my legs and put my feet in stirrups.

"There was a nurse holding my arms above my head so I wouldn't move.

"When they grabbed my testicles, they pinched them up. They took a needle and stuck it into my testicles." Hudlow believes this was anesthesia.

"They didn't wait for it to work. They made an incision. They went right on in there. I was hollering and crying. I was hurting."

None of the colony medical staff explained what they were doing to him, Hudlow said. "The only way I found out, an employee on Ward 7 told me I wouldn't be able to father any children.

"They treated us just like hogs, like we had no feelings."

Hudlow was released from the colony in October 1943 and drafted into the Army two months later.

"I went in at Omaha Beach in France in August 1944," two months after the Allied invasion of Europe.

Hudlow served as the radioman for his platoon leader. He saw combat in France, Belgium and Holland, where he was wounded in the left knee and captured by the Germans. He was in various prison camps for seven months until he was liberated by the Russians in May 1945.

Hudlow decided to make the military a career, serving 21 years in the Army and Air Force.

Hudlow doesn't talk about his war service unless questioned about it and did not mention his medals until asked if he was awarded any. He keeps his medals, citations and military records in a footlocker in his bedroom closet.

He said he has had more flashbacks about the sterilization procedure than about the terror of combat and imprisonment by the Germans.

"I remember this just as it was yesterday. It has always been in my mind. It has never left me."

He said his inability to have children "worked on my mind,

especially when I was around my sisters and my brother. They had children."[1]

Hudlow is not the only victim outraged by the wrong done to him. Fred Aslin and his eight brothers and sisters are also angry. After their father died during the Great Depression of the 1930s, they were taken from their mother who was unable to care for them and placed in the Lapeer State School, a closed psychiatric facility in Michigan hundreds of miles from their home. When they turned 18, the state sterilized them, one by one, against their will.

Michigan's law called for a hearing before a person could be sterilized. Fred Aslin was not permitted to attend his hearing. He had no attorney. Instead a guardian whom he never met represented him. At the hearing, a probate judge signed the papers for forced sterilization without comment and in August, 1944, the boy underwent surgery. In 1996, Aslin used the Public Information Act to find out why he was singled out. As he read the files kept by Lapeer, he learned for the first time how the authorities had justified the sterilization: "They termed us feeble-minded idiots, and wrote that our children would be like us or even worse." Yet neither he nor any of his siblings is mentally retarded. "My brother John always thought it was because we were just poor Indians," said Aslin, who is of mixed Ottawa and Chippewa ancestry.

Aslin demanded an apology from the state. When no apology came, he hired a lawyer and filed suit against the state of Michigan, seeking compensation. His claim was turned down because the statute of limitations had expired. However, one state official did send Aslin a letter of apology. After meeting with him, James K. Haveman, director of the Michigan Department of Community Health, wrote, "I am saddened that it took so long and so many had to suffer before the medical profession and judicial system realized how offensive the practice of sterilization was."[2]

Virginia and Michigan are not the only states that performed "eugenic sterilizations." There were victims in 28 other states as Becca Tanner, a reporter for the *Wichita* (Kansas) *Eagle*, discovered when she investigated the history of forced sterilization in Kansas. She discovered that the victims were "people with epilepsy, non-English-speaking immigrants, teenage girls who may have been raped or were pregnant out of wedlock, people suffering from depression or some form of mental illness, gays and lesbians, and, most frequently, criminals." Local officials tried to explain:

> "The whole idea of perfection of mankind meant that you were willing to experiment through science or social reform to do whatever you could to bring about a society that was better," said Virgil Dean, research historian at the Kansas State Historical Society in Topeka.

At one time, Kansas ranked third nationally in the number of sterilizations. The procedures were phased out in 1952, but the law allowing them remained on the books until the 1970s.

"There was a period of time when people thought this was the thing to do," said former Kansas Secretary of Social Rehabilitation Services Robert Harder. Harder, who served in that position from 1973 to 1987, said he was one of the people who finally insisted that the forced-sterilization law be repealed. "By then, we had begun to develop a more humane understanding of people and viewed sterilization as an inhumane practice," Harder said.

The effort to sterilize the unfit in Kansas began in 1894 with F. Hoyt Pilcher, then superintendent of Winfield's Kansas State Asylum for Idiotic and Imbecile Youth.

By 1895, Pilcher had developed a reputation as a trailblazer. The Winfield Courier reported: "The unsexing of one hundred and fifty of these inmates—male and female—was an innovation that received the endorsement of the entire medical profession of the world, and the plaudits of right thinking people everywhere."

Although forced sterilization had opponents from the start, by 1913 Kansas became one of the first states in the nation to pass a law saying forced sterilization was acceptable if "the mental or physical condition of any inmate would be improved . . . or that procreation by such inmates would be likely to result in defective or feebleminded children."

"Keep in mind, this was the same period that the same institutions were performing lobotomies," said Harder, the former SRS director. "Now, I don't think there is anyone around who would talk about lobotomy as an effective way or humane way to deal with persons with mental illness."

The Kansas sterilization law was declared constitutional in 1928 by the Kansas Supreme Court. The court relied on the US Supreme Court case of *Buck vs. Bell*, in which the court determined that "procreation of defective, feebleminded children with criminal tendencies does not advantage but patently disadvantages the race."

But as decades went by, people began raising questions.

A 1937 Time magazine story about the Girls Industrial School in Beloit led to the discovery that 62 of Beloit's inmates had been sterilized and 22 more had been scheduled to be sterilized.

A state investigation ensued, and the sterilizations were found to be illegal. An 18-year-old Beloit student quoted in the Kansas City Star following the investigation said: "All of us girls had been threatened before with sterilization unless we behaved ourselves. I knew it wouldn't do any good to kick although I didn't want it done. . . . I thought for a long while that life had very little left for me."

Dean, of the state historical society, said that even those who argued in favor of forced sterilization eventually began to see how it could be used on anybody—any group, any race. The American movement withered, particularly after World War II when news surfaced about how Nazis used eugenics to persecute people.[3]

CONNECTIONS

Why do you think state officials have been reluctant to apologize? What do they seem to fear? How does your answer explain the careful wording of the resolution passed by the Virginia House of Delegates? Why might it be easier for a state to express regret than to apologize?

Robert Harder told Becca Tanner that he was pleased that "light is now being shed on the forced sterilization laws in Kansas. It's like the stories about civil rights and how long it has taken for some of them to surface," he said. "For example, the Tulsa race riots were hardly talked about until recently. All this is just a reminder of the past and how some things were handled on the quiet side—and makes it difficult to track down."[4] What kinds of things are "handled on the quiet side"? Why is it important that "light" is shed on them?

In 2000, a number of groups in Virginia argued that making the Department for the Rights of Virginians with Disabilities independent of state government would be a good way to make amends to victims of forced sterilization. Do you agree? How might such a group have affected the outcome in Carrie Buck's case?

Find out about the history of the eugenics movement in your state. What efforts have been made to confront that history? To right old injustices? How important is it to acknowledge past wrongs even if they cannot be undone?

Why is it hard for public opinion to change? What is the role of education? Why is it important that a state or a nation set the record straight?

---

1. "Grim Legacy: Va. Eugenics Policy Led to Sterilization" by Bill Baskervill. The Associated Press, Feb. 6, 2001. Reprinted by permission of The Associated Press.
2. © 2000 The Washington Post Company.
3. "Eugenics Not Kansas' Proudest Moment" by Becca Tanner. *Wichita Eagle,* April 3, 2000.
4. Ibid.

7. Eugenics, Citizenship, and Immigration

America must be kept American. Biological laws show . . . that Nordics deteriorate when mixed with other races.

Calvin Coolidge

Between 1890 and 1914, over 15 million immigrants entered the United States. In some large cities, one out of every three residents was foreign-born. Many Americans felt threatened by the newcomers. In the early 1900s, economist Simon Patten described the way those fears were shaping American life:

> Each class or section of the nation is becoming conscious of an opposition between its standards and the activities and tendencies of some less-developed class. The South has its Negro, the city has its slums. . . . The friends of American institutions fear the ignorant immigrant, and the workingman dislikes the Chinese. Every one is beginning to differentiate those with proper qualifications for citizenship from some other class or classes which he wishes to restrain or exclude from society.

President Calvin Coolidge shared that consciousness. His concerns and those of other Americans about the effects of "race mixing" were heightened by eugenicists like Harry Laughlin and Carl Brigham (Chapter 5). They insisted that "according to all evidence available," "American intelligence is declining, and will proceed with an accelerating rate." They attributed the decline to the "presence here" of "inferior races." These eugenicists insisted that the nation could reverse the decline through laws that would "insure a continuously progressive upward evolution." They urged that those steps "be dictated by science and not by political expediency. Immigration should not only be restrictive but highly selective."

Brigham's *A Study of American Intelligence* and other books like it gave many Americans, including the president, a "scientific rationale" for their prejudices. These books also raised important questions about membership in American society. Who should be allowed to settle in the nation? What are "the proper qualifications for citizenship"? Chapter 7 explores the impact of the eugenics movement on the way ordinary Americans and their leaders answered these questions in the early 1900s. It also considers the consequences of those decisions on the lives of real people then and now. Like earlier chapters, Chapter 7 serves as reminder that science, in the words of physicist Leon M. Lederman, "can be used to raise mankind to new heights or literally to destroy the planet . . . . We give you a powerful engine. You steer the ship."

Guarded Gates or an Open Door?

Reading 1

In 1876, the United States celebrated the 100th anniversary of the Declaration of Independence. In honor of the event, the French gave the nation a huge copper statue that depicts liberty as a woman holding high a giant torch. Emma Lazarus, a Jew whose family had lived in the nation for generations, later wrote a poem describing the statue.

> A mighty woman with a torch, whose flame
> Is the imprisoned lightning, and her name
> Mother of Exiles. From her beacon-hand
> Glows world-wide welcome. . . .
> "Keep, ancient lands, your storied pomp!" cries she
> With silent lips.
> "Give me your tired, your poor,
> Your huddled masses yearning to breathe free,
> The wretched refuse of your teeming shore.
> Send these, the homeless, tempest tossed, to me.
> I lift my lamp beside the golden door."

In 1903, the year that Lazarus's poem was carved into the base of the Statue of Liberty, 10 percent of the nation was foreign-born. As immigration increased so did the fears of many native-born Americans. Native-born workers often viewed the newcomers as competitors for jobs, housing, and public services. More prosperous Americans felt threatened by the way the immigrants crowded into the nation's largest cities. Their legitimate concerns about the ability of local governments to deal with overcrowding turned into fears about the character of the newcomers. It was as if the new arrivals were the carriers of social problems rather than individuals who experienced those problems.

Like Emma Lazarus, Thomas Bailey Aldrich, came from a family that had lived in the United States for generations. He modeled his poem after the one she wrote, but the sentiment was very different. "The Unguarded Gate" was published in the *Atlantic Monthly*, the magazine he edited, in 1892.

> Wide open and unguarded stand our gates,
> Named of the four winds, North, South, East and West;
> Portals that lead to an enchanted land
> Of cities, forests, fields of living gold,
> Vast prairies, lordly summits touched with snow,

Majestic rivers sweeping proudly past
The Arab's date-palm and the Norsemen's pine—
A realm wherein are fruits of every zone,
Airs of all climes, for lo! throughout the year
The red rose blossoms somewhere—a rich land,
A later Eden planted in the wilds,
With not an inch of earth within its bound
But if a slave's foot press it sets him free!
Here, it is written, Toil shall have its wage,
And Honor honor, and the humblest man
Stand level with the highest in the law.
Of such a land have men in dungeons dreamed,
And with the vision brightening in their eyes
Gone smiling to the fagot and the sword.

Wide open and unguarded stand our gates,
And through them presses a wild motley throng—
Men from the Volga and the Tartar steppes,
Featureless figures of the Hoang-Ho,
Malayan, Scythian, Teuton, Kelt, and Slav,
Fleeing the Old World's poverty and scorn;
These bringing with them unknown gods and rites,
Those, tiger passions, here to stretch their claws.
In street and alley what strange tongues are these,
Accents of menace alien to our air,
Voices that once the tower of Babel knew!

O Liberty, white Goddess! is it well
To leave the gates unguarded? On thy breast
Fold Sorrow's children, soothe the hurts of fate,
Lift the down-trodden, but with hand of steel
Stay those who to thy sacred portals come
To waste the gifts of freedom. Have a care
Lest from thy brow the clustered stars be torn
And trampled in the dust. For so of old
The thronging Goth and Vandal trampled Rome,
And where the temples of the Caesars stood
The lean wolf unmolested made her lair.

In 1905, Francis Sargent, the commissioner general of immigration, was interviewed for a *New York Times* article entitled "Are We Facing an Immigration Peril?" He told a reporter:

"Put me down in the beginning as being fairly and unalterably opposed to what has been called the open door, for the time has come when every American citizen who is ambitious for the national future must regard with grave misgiving the mighty tide of immigration that, unless something is done, will soon poison or at least pollute the very fountainhead of American life and progress. Big as we are and blessed with an iron constitution, we cannot safely swallow such an endless-course dinner, so to say, without getting indigestion and perhaps national appendicitis."

"Do you mean that the danger is immediate or prospective?" he was asked.

"Both," he replied promptly. "Today there is an enormous alien population in our larger cities which is breeding crime and disease all the more dangerous because it is more or less hidden and insidious. But the greatest source of uneasiness has to do with the future. Under present conditions nearly one-half the immigrants who pass through [Ellis Island, the main port of entry for European immigrants] never get beyond New York City and State, or the immediately contiguous territory. Unless something is done to discourage this gradual consolidation, it is my fear and belief that within five years the alien population of the country will constitute a downright peril. . . ."

"During the past year there has been a notable increase in the number of criminals coming over here," [Sargent] continued, "some of them being the worst criminals in Europe. There is no question about it, for we have positive evidence of the fact. In short, the time has come for the country to demand to know the character of immigrants that Europe is shedding or trying to shed."

Continuing, the Commissioner stated that in several European cities, with or without the connivance of the authorities, inmates of hospitals and almshouses were, there was reason to believe, being provided with tickets and means of reaching Ellis Island.

Approximately 5 percent of deportation cases come under this class, he estimated.[1]

CONNECTIONS

In this reading three Americans who lived at the turn of the 20th century express their views of immigrants. List in your journal the adjectives each uses to describe immigrants. What images do these adjectives evoke?

What part do fears play in the way we perceive others? What is an alien? What is the difference between an immigrant and an alien? If the United States is a country of immigrants, are we all aliens?

Scapegoating is the practice of shifting blame and responsibility for a real or perceived failure from oneself to another individual or group. To what extent does each writer view immigrants as scapegoats responsible for all of society's ills? To what extent might the twisted science of eugenics provide a rationale for the practice of scapegoating?

Although neither Aldrich nor Sargent uses the word *eugenics*, how are the concerns they express similar to those of Francis Galton and Charles Davenport? (See Chapter 3.) On what issues do you think Sargent, Aldrich, and Davenport might agree? Where might they differ?

Modern historians and economists note that immigrants in the early 1900s were as skilled and well educated as most Americans of their day. Although many were unable to read or write, so were many Americans. Sargent and others who opposed immigration often compared immigrants as a group to Americans as a nation. But nearly 80 percent of the immigrants were between the ages of 16 and 44 and about 70 percent were men. If opponents of immigration had compared the newcomers to a group of Americans in the same age range and with a similar gender balance, they would have found the two groups more alike than different. How does the way we use numbers shape the way we define an issue? The conclusions we reach? What other factors may affect the way we define an issue like immigration?

---

1. *The New York Times*, January 29, 1905, pp. 26, 28.

Reading 2

In the early 1900s, many Americans saw immigrants as the "other"—people inherently unlike *us*. They focused on differences in clothing, language, and customs and ignored similarities. Many of them never knew the newcomers as individuals—as people with hopes and dreams similar to their own.

In the 1970s, Demetrius Paleologas, a Greek immigrant, recalled how he looked when he arrived in the United States in 1915 at the age of nineteen.

> I came to St. Louis, to my father's friend. He says, "I'll take you in." If I tell you the condition we were in—lice—oh, you have no idea. So he took me to a clothing store and he bought me underwear, socks, shoes, whole suit of clothes, shirt, and everything. And he took me to his place of business—he had a small restaurant—and they had a shower downstairs. He said, "Take all your clothes, throw them down there, wash yourself good, and put the new clothes on."
>
> This man was very nice and he gave me a job in his restaurant—wash dishes. We used to live with three, five, six beds in one room, over the restaurant. Then immediately I thought that I should learn how to speak and how to write, learn the language. Not only that, but I says, "Where am I going to go now? Remain a dishwasher all the time? That's no good. I don't like to remain a dishwasher." And after I was doing the dishes, I was looking at the cooks, and I tried to help the cooks. And in the evening—seven o'clock in the evening—I walk about a mile and a half, walk like the dickens, to go down to the Lincoln Avenue School and start learning the English language.
>
> In six months, I became a third cook, then I became a second cook. Inside a year, one of the chef happen to be sick and I took over as a chef, too. But I said to myself, "I'm going to become a cook, how much I'm going to make?" So I ask the floor boss, "I want to come into the dining room and help—you know, the busboys and like that. Could you give me a job?" So he give me a job.
>
> In 1920—almost five years later—I decide to go into business for myself.[1]

How typical was Paleologas's experiences? Historian Steven J. Diner tries to set experiences like those of Paleologas's in a larger context:

Immigrants awaiting the ferry to Ellis Island at the turn of the 20th century.

Most of the immigrants who came to America between 1890 and World War I sought economic opportunity more than personal liberty; many intended to return home once they earned some money. Most immigrants although poor did not come from the poorest of the poor, and few lacked homes. Emigration cost money, a carefully calculated investment enabling the sojourners to earn in America the funds needed to increase their modest landholdings and possessions back home. They could hardly be described as tired. Young, ambitious, and accustomed to hard work, immigrants acted boldly and deliberately to gain control over their lives. These artisans and farmers, refusing to accept passively the negative effects of industrial capitalism in their homelands, came to America to find economic security for their families.

More immigrants arrived during the Progressive Era (1890–1914) than ever before or after, fifteen million in the twenty-four years between 1890 and 1914, although the foreign-born proportion of the US population remained nearly the same in 1910 (14.5 percent) as in 1860 (13.2 percent). The sources of immigration changed substantially, however. Before 1890, most immigrants had come from Great Britain, Ireland, Canada, Germany, Scandinavia, Switzerland, and Holland. Immigrants after 1890 came disproportionately from the Austro-Hungarian Empire, Italy, Russia, Greece, Romania, and Turkey. Eighty-seven percent in 1882 arrived from the countries of Northwestern Europe, but by 1907, 81 percent hailed from the South and East. A majority of the "new" immigrants were not Protestants, and they spoke languages, such as Polish, Yiddish,

Lithuanian, Czech, and Greek, that were completely unfamiliar to Americans.

To be sure, immigrants continued to come to America from Northwestern Europe. Between 1890 and 1920, 874,000 people entered from Ireland, 991,000 from Germany, 571,000 from Sweden, 352,000 from Norway, but they drew little attention when compared with the 3,807,000 from Italy, for example. Substantial numbers also came from outside Europe, particularly from French and English Canada, Japan (until excluded by diplomatic agreement in 1906), Mexico, and Syria.[2]

CONNECTIONS

Create an identity chart for Paleologas. What does he add to our understanding of what it was like to be an immigrant in the early 1900s? How does his story challenge the way Charles Davenport and other eugenicists viewed "the immigrant" (Chapter 3)? The views expressed by Thomas Aldrich and Francis Sargent in the previous reading?

What is the meaning of the word *assimilation*? To what extent did Paleologas become assimilated? What does an immigrant give up when he or she becomes assimilated? What does he or she gain?

Compare the list of adjectives you compiled in the previous reading with Paleologas's experiences. What similarities do you notice? How do you account for differences?

Why do you think the man who took Paleologas in was able to see beyond the dirt and the lice? What attitudes and values make it possible for someone to see beyond outward appearances? To know another person as an individual rather than as a stereotype?

How do you think Paleologas would have responded to Francis Sargent's remarks (Reading 1)? What would he want Sargent to know about him and his fellow immigrants?

---

1. "Demetrius Paleologas" in *American Mosaic* by Joan Morrison and Charlotte Fox Zabusky. New American Library, 1980, p. 75.
2. Excerpt from *A Very Different Age: Americans of the Progressive Era* by Steven J. Diner. Copyright © 1998 by Steven J. Diner. Reprinted by permission of Hill and Wang, a division of Farrar, Straus, and Girous, LLC, pp. 76–77.

Reading 3

The nation's lawmakers decide who may settle in the United States. Every immigration law excludes, distinguishes, or discriminates based on real or imagined differences. The chart below outlines changes in American immigration policies.

U.S. Immigration Policy 1789–1920

1789–1875	Everyone
1875	No convicts No prostitutes
1882	No idiots No lunatics No one requiring public care No person who cannot pay a head tax of 50 cents
1882–1943	No Chinese
1885	No cheap contract laborers
1891	No immigrants with contagious diseases No paupers No polygamists (Start of medical inspection)
1903	No epileptics No insane persons No beggars No anarchists
1907	No feebleminded No children under 16 unaccompanied by parents No immigrants unable to support themselves because of physical or mental defects
1917	No immigrants from most of Asia or the Pacific Islands No illiterate adults (start of literacy tests)

CONNECTIONS

"Who am I?" is a question almost everyone asks at one time or another. In answering, we define ourselves. Nations, like individuals, have an identity. Add to the identity chart you created in Chapter 4 for the United States in the early 1900s based on information provided in this reading. Begin with the words or phrases that Americans used to describe themselves. Then add the labels others might have attached to the nation. What does the chart provided in this reading add to your understanding of American identity in the 1800s and early 1900s?

A nation's identity—its sense of who it is and what it might become—is more than a set of labels. It is also shaped by a philosophy—the ideas, values, and beliefs that affect the way its people understand the world and their place in the world. What ideas about the United States and its place in the world does the chart suggest? What does it suggest about how the nation's philosophy had evolved since its founding?

Sociologist Kai Erikson has noted that one of the surest ways to "confirm an identity, for communities as well as for individuals, is to find some way of measuring what one is not." What individuals and groups were not included in the word *American* in the 1800s and early 1900s? Who is not included in the word today? What did it mean to be excluded in the nineteenth and early twentieth centuries? What does it mean today?

Reading 4

Over the years, the U.S. Congress has considered and reconsidered not only the question of who may settle in the United States but also who is entitled to citizenship. In 1790, the nation's lawmakers offered citizenship to "the worthy part of mankind." To become a citizen, an immigrant had to live in the United States for two years and provide proof of good character in court. Immigrants also had to be white. Non-whites could live in the nation but could not become citizens, even though their American-born children were citizens by birth.

After the Civil War, Senator Charles Sumner of Massachusetts suggested that "all acts of Congress relating to naturalization be . . . amended by striking the word 'white' wherever it occurs, so that in naturalization there still be no distinction of race or color." He encountered immediate objections from western senators. Historian Matthew Frye Jacobson writes:

> Both the significance of Sumner's proposal and the ramifications of Western dissent were acknowledged and summed up in an amendment proposed by a . . . senator in jest, "Provided, that the provisions of this act shall not apply to persons born in Asia, Africa, or any of the islands in the Pacific, nor to Indians born in the wilderness. [Laughter]". . . .
>
> Sumner himself announced that, in striking the word "white," he merely wanted to "bring our system in harmony with the Declaration of Independence and the Constitution of the United States." "The word 'white,'" he offered, "cannot be found in either of these great title-deeds of this Republic." To senators from the West, by contrast, the word provided a critical bulwark against national decline. "Does the Declaration mean," one wanted to know, "that the Chinese coolies, that the Bushmen of South Africa, that the Hottentots, the Digger Indians, heathen, pagan, and cannibal, shall have equal political rights under this Government with citizens of the United States?" The implicit logic of this list is telling in its very confusion. "White," by implication here, is a designation that indicates not only color but degree of freedom (as against "coolies"), level of "civilization" (as against "cannibals"), and devotion to Christianity (as against "pagans" and "heathens").[1]

In the end, Congress decided to keep the word *white* and add to those eligible for citizenship persons "of the African race or of African descent." The change failed to address an important question: Who is white? It was a question that would be raised in the nation's courts for years to come. The first person to do so was a Chinese immigrant named Ah Yup. In 1878, he asked the court whether a person of the "Mongolian race" qualified as a "white person." The judge replied:

> The words white person. . . constitute a very indefinite description of a class of persons, where none can be said to be literally white, and those called white may be found of every shade from the lightest blonde to the most swarthy brunette. But these words in this country, at least, have undoubtedly acquired a well settled meaning in common popular speech and they are constantly used in the sense so acquired in the literature of the country, as well as in common parlance. As ordinarily used everywhere in the United States, one would scarcely fail to understand that the party employing the words "white person" would intend a person of the Caucasian race.[2]

The judge went on to quote Johann Blumenbach and other scholars (Chapter 2). Despite their lack of agreement, he wrote, "No one includes the white, or Caucasian, with the Mongolian or 'yellow race' and no one of those classifications recognizing color as one of the distinguishing characteristics includes the Mongolian in the white or whitish race." The ruling raised a new question: What is a *whitish race*? In the years that followed, the struggle to define "whiteness" continued. Were Armenians white? Hawaiians? Syrians? The Burmese? Turks? Are people from India white? What about Mexicans? In each case, judges relied on a combination of "race science," eugenics, and "popular understanding" to determine who was "white." Some even consulted segregation laws to determine who was white. A number of these laws noted that anyone who was not legally black was "white."

Two cases in the early 1920s illustrate how race was used to guard the privileges of white Americans. In October 1922, Takao Ozawa, an immigrant from Japan, petitioned the courts for the right to become a U.S. citizen. He argued that the 1875 law that extended citizenship to "Africans" was inclusive rather than exclusive. He noted that Congress had passed a law in 1882 that barred the Chinese from settling in the United States for ten years. From time to time that law was renewed without a single mention of Japanese immigrants. He also cited cases where judges had ruled that anyone not black was "white." And finally, he observed, "The Japanese are 'free.' They, or at least the dominant strains, are 'white persons,' speaking an Aryan tongue and having Caucasian root stocks; a superior class, fit for citizenship. They are assimilable."[3]

The US Supreme Court ruled against Ozawa, arguing that he was "white" but not "Caucasian." In 1923, just a few months after the Ozawa decision, a similar case reached the Supreme Court. This time the government wanted to take away citizenship from Singh Thind, a Hindu from India, because he was "not white." This time, the same justices who denied Ozawa citizenship because he was "white" but not "Caucasian" ruled that Thind was also ineligible because he was "Caucasian" but not "white." They stated, "It may be true that the blond Scandinavian and the brown Hindu have a common ancestor in the dim reaches of antiquity, but the average man knows perfectly well that there are unmistakable and profound differences among them today."[4]

Although judges continued to quote "race scientists," eugenicists, and anthropologists, they clearly saw race not as a matter of science but as a "practical line of separation" among the peoples of the world. How they drew that line varied case by case, incident by incident.

CONNECTIONS

Each of us has a "universe of obligation." Whom did Sumner consider a part of his universe of obligation? His "moral community"? How did other senators define their "moral community"? What were the consequences of the way they defined the nation's universe of obligation?

Whom did the judges regard as assimilable? Create a working definition of the word *assimilate*. Include your own understanding as a well the way a dictionary defines the term and the meanings attached to it in this reading. How is the word *assimilate* related to the way the Senate defined the word *white*? What do you think the word *whitish* means? What is the significance of the term?

Applicants for citizenship in the early 1900s were all men. A woman derived her citizenship from her father and later her husband. A law passed in 1907 stated that any woman born in the United States who married a citizen of another country would lose her citizenship. The law remained on the books until 1922, when Congress separated a woman's citizenship from that of her husband. How did the 1907 law regard women? What fears did it address? On what values was it based?

What does it mean to see race as a matter of science? As a "practical line of separation"? Who draws the line in either case? For whom is that line "practical"? In reflecting on divisions in American society, sociologist David Schoem writes:

> The effort it takes for us to know so little about one another across racial and ethnic groups is truly remarkable. That we can live

> so closely together, that our lives can be so intertwined socially, economically, and politically, and that we can spend so many years of study in grade school and even in higher education and yet still manage to be ignorant of one another is clear testimony to the deep-seated roots of this human and national tragedy. What we do learn along the way is to place heavy reliance on stereotypes, gossip, rumor, and fear to shape our lack of knowledge.[5]

To what extent does the debate in Congress after the Civil War support Schoem's observations? Use newspapers, magazines, and other media to find out the extent to which current debates in Congress support Schoem's view?

---

1. Reprinted by permssion of the publisher from *Whiteness of a Different Color: European Immigrants and the Alchemy of Race* by Matthew Frye Jacobson, Cambridge, Mass: Harvard University Press, Copyright © 1998 by the President and Fellows of Harvard College.
2. Ibid., p. 227.
3. Ibid., p. 234.
4. Ibid., p. 236.
5. *Separate Worlds* by David Schoem. University of Michigan Press, 1991, p. 3.

Reading 5

Every debate, including the ones over immigration and naturalization, takes place within a context. The debates over immigration and naturalization reflected the fears and concerns of many Americans about differences and membership. They also reflected a belief that the world was a very dangerous place. World War I (1914–1918) intensified that belief.

Before World War I, it was possible to travel anywhere in the world without a passport or visa. Wartime fears of spies and anxieties over open borders changed the way nations regarded not only immigrants but also tourists and business travelers. Those fears and anxieties remained after the war ended. For many Americans, the most visible sign of danger was Ellis Island, where record-breaking numbers of immigrants were arriving daily. Many were among the millions of refugees forced from their homelands by war, revolution, and government decree in the years after the war. Columnist Dorothy Thompson described them this way:

> A whole nation of people, although they come from many nations, wanders the world, homeless except for refuges, which may at any moment prove to be temporary. They are men and women who often have no passports; who, if they have money, cannot command it; who, though they have skills, are not allowed to use them. This migration—unprecedented in modern times, set loose by the World War and the revolutions in its wake—includes people of every race and every social class, every trade and every profession.[1]

In the United States, a story in *The New York Times* in 1920 about some of those refugees created an uproar. On August 17, *The Times* reported:

> Leon Kamaiky, [a commisioner of the Hebrew Immigrant Aid Society (HIAS) and] publisher of the *Jewish Daily News* of this city returned recently from Europe, where he went together with Jacob Massel, to bring about the reunion of Jewish families who were separated by the war. Mr. Kamaiky has been abroad since last February. . . .
>
> In an article in the *Jewish Daily News* describing conditions in Eastern Europe, Mr. Kamaiky declared that "if there were in existence a ship that could hold 3,000,000 human beings, the 3,000,000 Jews of Poland would board it and escape to America."

Alarmed readers wanted to know if this meant that the HIAS was planning to bring over three million Polish Jews. Members of Congress responded by calling for a ban on all immigration for a period of time—some favored a six-month ban, while Representative Albert Johnson, the chairman of the House Committee on Immigration, called for a two-year ban. He argued that "the new immigration is not the kind or quality to meet the real needs of the country. We are being made a dumping ground. We are receiving the dependents, the human wreckage of the war, not the strength and virility that once came to hew our forests and till our soil." He brought his bill to a vote without a single hearing. After some negotiating, the ban was reduced to one year and the House quickly passed Johnson's bill.

The Senate Committee on Immigration was more cautious. Its chair told reporters, "This talk about 15,000,000 immigrants flooding into the United States is hysteria and not based on actual information." He then called for hearings on the bill. The first witness was Johnson who presented a report "confirming the statement if there were in existence a ship that would hold 3,000,000 human beings, the 3,000,000 Jews of Poland would board it to escape to America." He warned that unless an emergency act was passed, European immigration would "flood this country as soon as the war passport system went out of existence."

When John L. Bernstein, the president of HIAS, was called to testify, he tried to clarify the situation. He told senators that the rumors were false. HIAS had no plans to bring three million Polish Jews to the United States. The group was not even planning to send, as the American consul in Poland claimed, "250,000 emigrants of one race alone, the Jewish, to the United States within the next three years." He bluntly stated:

> Now, gentlemen, . . . our most prosperous year was the year 1919. . . . During the year 1919 we obtained the largest contributions, both in membership and in donations, we have ever received, . . . and the amount of the contributions was $325,000. . . .
>
> Now, I will leave it to you, gentlemen, how much of that $325,000 will be left us to undertake this great plan that somebody is reading, about the bringing over of 250,000 emigrants here?[2]

Senator Hiram Johnson of California asked Bernstein whether HIAS encouraged or discouraged immigration. Bernstein replied, "Well, to be perfectly frank, we do neither. A man comes to our office for advice; we give it to him. And remember, we do not come in contact with any person unless he is already an emigrant, because we have no offices throughout Europe. . . . Our work in Poland is merely police work. We are trying to prevent the emigrants in Poland from being exploited, cheated, and swindled."[3]

In the end the Senate decided that there was no emergency nor were there grounds for a general ban on immigration. Still, like their counterparts in the House of Representatives, many senators were uneasy about the "quantity" and "quality" of the nation's newest arrivals. In 1921, the House and the Senate passed the first of several laws limiting immigration.

CONNECTIONS

How do you account for the hysteria that resulted from a brief story in *The New York Times*? What fears fueled the hysteria? What prejudices heightened those fears? Why do you think fears related to immigration tend to increase in war time?

Dorothy Thompson believed that no democratic nation can "wash its hands of [the problems of the refugees] if it wishes to retain its own soul." How do you think a lawmaker like Albert Johnson would respond to her statement? How might a eugenicist like Harry Laughlin respond?

Thompson insisted that "democracy cannot survive" if people deny minorities "the right to existence." How does she define the word *democracy*? Why does she believe that a democracy must protect the rights of minorities? Do you agree? Would Johnson agree?

---

1. *Refugees: Anarchy or Organization?* by Dorothy Thompson. Random House, 1939, p.1.
2. Quoted in *Shores of Refuge* by Ronald Sanders. Henry Holt and Co., 1988, p. 385.
3. Ibid., p. 385.

Reading 6

Efforts to control immigration had the support of many Americans. As early as 1894, a number of graduates of Harvard University openly expressed their fears of the "inferior hordes of degenerate peoples" who were crowding into the nation. That year they founded the Immigration Restriction League. Their first political victory came during World War I. In 1917, they persuaded Congress to enact a bill requiring that every immigrant pass a literacy test.

The literacy test was just the beginning. Members of the League wanted further restrictions. To make their case, they relied on statistics from the Eugenics Record Office and the organizational abilities of Harry Laughlin. They persuaded the House Committee on Immigration to hold hearings on the "immigration problem" in 1920. The committee consisted of 15 members of the House of Representatives and was chaired by Albert Johnson. Johnson was so impressed with Harry Laughlin's testimony that he appointed Laughlin "Expert Eugenics Agent of the House Committee on Immigration."

Harry Laughlin.

Whenever Laughlin testified, he brought graphs, pedigree charts, and the results of hundreds of IQ tests that were administered to soldiers during World War I as evidence of "the immigrant menace." At one hearing, he plastered the walls of the meeting room with photographs taken at Ellis Island. Above the photos hung a banner that read "Carriers of the Germ Plasm of the Future American Population."[1] Laughlin told committee members:

> The matter of social and cultural assimilation of immigrants has just come to an acute state in the United States. The formation of isolated alien centers, which maintain their alien languages and cultures, is a dangerous thing for the American people. . . .
>
> If the American Nation decides that it is still unmade as a people, then it might as well throw open the doors and admit all comers, but if it decides that we have national ideals worth saving, not only in national tradition and individual quality, but also racial ingredients, the Nation must exercise stricter control over immigration. This is a critical period in American history. We can continue to be American,

> to recruit to and develop our racial qualities, or we can allow ourselves to be supplanted by other racial stocks.[2]

Scientists who publicly disputed Laughlin's findings were ignored. For example, when Herbert Spencer Jennings, a former eugenicist and a respected biologist, told the committee that Laughlin's statistics were flawed, his testimony was cut short. Members of Congress were not interested in hearing that Laughlin's charts and graphs proved the opposite of what he claimed that they proved. Most newspapers and magazines also ignored Jennings's testimony. Reporters found Laughlin's lurid findings more compelling. After all, those findings confirmed what many Americans already believed: immigrants were "different" and those differences threatened the American way of life.

CONNECTIONS

Why do you think Americans paid more attention to the eugenicists than to their critics? What was the appeal of an exclusive rather than an inclusive nation?

Why did Harry Laughlin and members of the Immigration Restriction League believe the recent immigrants would never become socially or culturally assimilated? How does he seem to define the word *assimilate*? How do you define it? To what extent were the immigrants quoted in Chapter 4 assimilated? What might they have added to the picture Harry Laughlin painted?

What does Laughlin mean when he says, "We can continue to be American, to recruit to and develop our racial qualities, or we can allow ourselves to be supplanted by other racial stocks." How does he seem to define the word *American*? President Calvin Coolidge supported restrictions on immigration because "America must be kept American." How did he seem to define the word? Look carefully at the quotations that follow. How does each writer define the word *American*? Which definitions are closest to those of Coolidge and Laughlin? To the views of Emma Lazarus (Reading 1)? Which are closest to the way you define the word?

> —In 1782, French immigrant Jean de Crevecoeur wrote, "He is an American who, leaving behind him all his ancient prejudices and manners, receives new ones from the new mode of life he has embraced, the new government he obeys, and the new rank he holds."

> —In the 1850s, Theodore Parker, a minister of British descent, argued that an American is someone who believes "not 'I am as good as

you are' but 'You are as good as I am.'"

—In the 1920s, Boston Mayor James Michael Curley, an Irish American, stated, "All of us under the Constitution are guaranteed equality, without regard to race, creed, or color. If the Jew is barred today, the Italian will be tomorrow, then the Spaniard and Pole, and at some future date the Irish."

—In 1939, newspaper columnist Dorothy Thompson, the daughter of an English immigrant wrote, "George Washington was only born in this country because his grandfather was a political refugee. William Penn fled to this country from the prisons of England, where his fight for freedom of conscience . . . kept him continually locked in various jails. Thomas Paine may be called the original author of the Declaration of Independence, and he was twice a refugee of this country—once from the conservatism of England and once from the terror of the French Revolution. Woodrow Wilson's forebears were religious refugees from Ireland; the LaFollette family were Huguenot refugees; the Middle West was settled to its great advantage by many Forty-Eighters [refugees from the Revolution of 1848 in Germany], and among those Forty-Eighters was the father of Justice [Louis] Brandeis and the father of Adolph Ochs [the publisher of *The New York Times*].

—In 1949, Langston Hughes, a noted African American poet, wrote:
Oh, yes,
I say it plain,
America never was America to me.
And yet I swear this other—
America will be!

---

1. Frances Hassencahl, "Harry H. Laughlin, 'Expert Eugenics Agent for the House Committee on Immigration and Naturalization, 1921 to 1931." UMI Dissertation Services, 1970, p. 247.
2. House Committee on Immigration Hearings, "Europe as an Emigrant Exporting Continent and the United States as an Immigrant Receiving Nation," March 8, 1924, pp.1294–1295.

Reading 7

In his testimony before the House Committee on Immigration, John Trevor, a New York attorney and member of a group called the Allied Patriotic Societies, proposed that Congress limit immigration country by country to two percent of the immigrants from that country living in the United States in 1890. The date was critical, because most immigrants from southern and eastern Europe arrived after 1890. The House of Representatives debated Trevor's plan in March and April of 1924. Excerpts from the debate reveal how strongly members felt about immigration. It also reveals the extent of Harry Laughlin's influence.

Representative Clarence F. Lea of California told his fellow lawmakers:

> What is that assimilation that we demand of a naturalized citizen? Assimilation requires adaptability, a compatibility to our Government, its institutions, and its customs; an assumption of the duties and an acceptance of the rights of an American citizen; a merger of alienism into Americanism.
>
> True assimilation requires racial compatibility. Nature's God has given the world a brown man, a yellow man, and a black man. Whether given to us by the wisdom of a Divine Ruler or by our own prejudices or wisdom we have a deep-seated aversion against racial amalgamation or general social equality with these races. Members of these races may have all the moral and intellectual qualities that adorn a man of the white race.
>
> Many individuals of any race may be superior, by every just standard of measurement, to many individuals of the white race. Yet there is an irreconcilable resistance to amalgamation and social equality that cannot be ignored. The fact is it forms an enduring barrier against complete assimilation. The brown man, the yellow man, or the black man who is an American citizen seeks the opportunities of this country with a handicap. It may be humiliating or unjust to him. You may contend it is not creditable to us, but it does exist. It causes irritation, racial prejudice, and animosities. It detracts from the harmony, unity, and solidarity of our citizenship.
>
> But to avoid further racial antipathies and incompatibility is the duty and opportunity of this Congress. The first great rule of exclusion should prohibit those non-assimilable. Our own interests, as well as the ultimate welfare of those we admit, justify us in prescribing a strict rule as to whom shall be assimilable. We should require physical,

moral, and mental qualities, capable of contributing to the welfare and advancement of our citizenship. Without these qualities it would be better for America that they should not come.

Representative Adolph J. Sabath of Illinois saw assimilation from a different perspective. He argued:

> What is meant by assimilation is difficult of definition. The mere fact that an immigrant, when he arrives or even after he has lived here for a number of years, still speaks his native language does not indicate that he is not being assimilated. Every day that he lives here he imbibes American ideas. . . .
>
> Whatever his garb may have been when he came, the first suit of clothes that he purchases with his honestly acquired earnings, which represent his creative efforts from which the country profits, is made according to the American model. His work is performed in accordance with the methods adopted in our industrial centers. He becomes familiar with our form of government. His acquaintance with our laws equals that of the average inhabitant of our country, and his obedience to them measures up to that of the average native. It is true that he reads books and newspapers printed in foreign languages, but it is by means of them that he acquires a fund of information relative to the true spirit of America. Anybody familiar with the foreign language press, and with what it has done in the direction of educating the immigrant into an appreciation of what America stands for, can testify to this fact. The children of these foreign parents brought up in American public schools grow up without even an ability to read the foreign press.
>
> The majority in its report . . . unjustifiably charged and contended that there is in this country an undigested mass of alien thought, alien sympathy, and alien purpose which creates alarm and apprehension and breeds racial hatreds. This, like most figures of speech, can not bear analysis. What is meant by alien thought and alien purpose as applied to immigrants? Does it mean that they are opposed to the land in which they live, in which they earn their livelihood, where they have established a permanent home for themselves and their children? Does it mean that they would invite conquest by foreign nations, and having to a great extent left the lands of their birth because deprived of liberty and that freedom which they enjoy in this country, that they would be willing to forego the blessings that have come to them under our benign institutions? Have they not by coming

here severed their political relations with foreign lands? Does any considerable portion of them ever expect to leave our shores? Have the thought and purpose of that Europe which they left behind been such as to attract instead of increase the repulsion which drove those immigrants to America? Are men apt to choose misery and unhappiness when they are enjoying contentment and comparative prosperity and are looked upon not as cannon fodder but as men? As well might it be said that the Puritans of New England, the Cavaliers of Virginia and Maryland, the Knickerbockers of New York, the Quakers of Pennsylvania, and the Scandinavians of the Middle West brought with them undigested masses of alien thought, alien sympathy, and alien purpose, which made of them a menace to this country.

> It is not the immigrants who are breeding racial hatreds. They are not the inventors of the new anthropology. Nor do they stimulate controversy. It would rather appear, in fact is clearly shown, to be those who are seeking to restrict or to prohibit immigration who entertain such sentiments and who are now attempting to formulate a policy which is, indeed, alien to the thought, the sympathy, and the purpose of the founders of the Republic and of that America which has become the greatest power for good on earth.

Representative Grant M. Hudson of Michigan took issue with the idea that immigrants change their customs and their attitudes. He told Congress:

> The "melting pot" has proved to be a myth. We are slowly awakening to the consciousness that education and environment do not fundamentally alter racial values.
>
> Today we face the serious problem of the maintenance of our historic republican institutions. Now, what do we find in all our large cities? Entire sections containing a population incapable of understanding our institutions, with no comprehension of our national ideals, and for the most part incapable of speaking the English language. Foreign language information service gives evidence that many southern Europeans resent as an unjust discrimination the quota laws and represent America as showing race hatred and unmindful of its mission to the world. The reverse is true. America's first duty is to those already within her own shores. An unrestricted immigration policy would work an injustice to all, which would fall hardest on those least able to combat it.
>
> George Washington in his Farewell Address said: Citizens by

birth or choice, of a common country, that country has a right to concentrate your affection. . . . [W]ith slight shades of difference, you have the same religion, manners, habits, and political principles.

Washington observed—slight shades of difference.

But today we see huge masses of non-American-minded individuals, living in colonies or ghettoes, or even cities and counties of their own. Here they perpetuate their racial mindedness, their racial character, and their racial habits. Here they speak their own tongue, read their own newspapers, maintain their separate educational system.

Ira Hersey of Maine offered his view of the nation's history:

Mr. Chairman, the New World was settled by the white race. True, we found here when the Pilgrim Fathers landed the red race. The Indian was never adapted to civilization. His home was the forest. He knew no government. He cared nothing for civilization. He gave freely of his land to the white man for trinkets to adorn his person; and this race of people, the first Americans, were pushed back as the forests receded until to-day he occupies here and there small portions of the United States, living the primitive life, wards of this Government, and in a few years they will be known no more forever.

They never were a menace to the Government. They have never been known in politics. On account of race and blood they have never been able to assimilate with our people and have kept their own place and have caused very little trouble in the progress of civilization in this country.

America! The United States! Bounded on the north by an English colony, on the south by the Tropics, and on the east and west by two great oceans, was, God-intended, I believe, to be the home of a great people. English speaking—a white race with great ideals, the Christian religion, one race, one country, and one destiny. [Applause.]

It was a mighty land settled by northern Europe from the United Kingdom, the Norsemen, and the Saxon, the peoples of a mixed blood. The African, the Orientals, the Mongolians, and all the yellow races of Europe, Asia and Africa should never have been allowed to people this great land.

Meyer Jacobstein of New York had a more expansive view of citizenship. He insisted:

Perhaps the chief argument expressed or implied by those favoring the Johnson bill [the National Origins Act] is that the new immigrant is not of a type that can be assimilated or that he will not carry on the best traditions of the founders of our Nation, but, on the contrary, is likely to fill our jails, our almshouses, and other institutions that impose a great tax burden on the Nation.

Based on this prejudice and dislike, there has grown up an almost fanatical anti-immigration sentiment. But this charge against the newcomers is denied, and substantial evidence has been brought to prove that they do not furnish a disproportionate share of the inmates of these institutions.

One of the purposes in shifting to the 1890 census is to reduce the number of undesirables and defectives in our institutions. In fact, this aspect of the question must have made a very deep impression on the committee because it crops out on every occasion. The committee has unquestionably been influenced by the conclusions drawn from a study made by Dr. Laughlin.

This is not the first time in American history that such an anti-foreign hysteria has swept the country. Reread your American histories. Go back and glance through *McMaster's History of the United States* covering the years from 1820 to 1850. You will find there many pages devoted to the "100 per centers" of that time. So strange was the movement against the foreigner in those decades before the Civil War that a national political party, the "Know-Nothing Party," sought to ride into power on the crest of this fanatical wave.

In those early days, however, the anti-foreign movement, strangely enough, was directed against the very people whom we now seek to prefer—the English, the Irish, and the Germans. The calamity howlers of a century ago prophesied that these foreigners would drag our Nation to destruction.

The trouble is that the committee is suffering from a delusion. It is carried away with the belief that there is such a thing as a Nordic race which possesses all the virtues, and in like manner creates the fiction of an inferior group of peoples, for which no name has been invented.

Nothing is more un-American. Nothing could be more dangerous, in a land the Constitution of which says that all men are created equal, than to write into our law a theory which puts one race above another, which stamps one group of people as superior and another as inferior. The fact that it is camouflaged in a maze of statistics will not protect this Nation from the evil consequences of such an unscientific, un-American, wicked philosophy.

In the end, the bill passed by an overwhelming majority in both the House of Representatives (373 to 71) and the Senate (62 to 6). In May 1925, President Calvin Coolidge signed the National Origins Act into law.

CONNECTIONS

A number of Congressmen quoted in this reading try to define the word *assimilate*. How do dictionaries define the word? What does the word mean to you? Why is the word so central to the debate?

Which representatives argue for immigration restriction? What do they fear? What do their speeches suggest about racial attitudes in the 1920s? About the influence of eugenics?

What points do Meyer Jacobstein and Adolph Sabath emphasize in their opposition to the bill? What do they fear? What do their speeches suggest about their racial attitudes? How does each representative define the word *American*? What do all five definitions have in common? On what points do they differ?

According to Sabath, who is breeding racial hatred? Why does he see their efforts as "alien to the thought, the sympathy, and the purpose of the founders of the Republic and of that America which has become the greatest power for good on earth"? How might a eugenicist respond to his attack?

The full text of the debates appears in the *Congressional Record for March and April 1924*, along with charts and graphs from Laughlin's exhibits. They can be used to prepare a report on regional voting patterns. Which regions of the country show the strongest support for the bill? Which show the least support? How do you explain the geographic division?

In the 1990s there were renewed calls for immigration restriction. Review newspaper and magazine articles on this topic. How do the recent arguments differ from those of the 1920's? How are the debates similar?

Reading 8

In 1924, President Calvin Coolidge told the American people, "Restricted immigration is not an offensive but purely a defensive action. It is not adopted in criticism of others in the slightest degree, but solely for the purpose of protecting ourselves. We cast no aspersions on any race or creed, but we must remember that every object of our institutions of society and government will fail unless America be kept American."

Coolidge's views were based on his understanding of eugenics and his belief in the racial superiority of "Caucasians." Many American voters as well as members of Congress shared those views. The new law was extremely popular. It seemed to solve the nation's "immigrant problem."

The people who opposed restrictions on immigration and deplored the language the eugenicists used to shape public opinion were those who saw the immigrants as individuals and understood their plight. One of those was Connie Young Yu's maternal grandmother. She was one of many Chinese women held at the Angel Island immigration station in San Francisco Bay in 1924. Yu's grandfather was born in the United States and was therefore an American citizen. So were his children. Although his wife was born in China, she too was a U.S. citizen according to American law at the time of her marriage. Yet when she and her young children tried to return to the United States from China after her husband's death, a health inspector said she had filariasis, liver fluke, "a common ailment of Asian immigrants which caused their deportation by countless numbers. The authorities thereby ordered Grandmother to be deported as well," writes Yu.

> While her distraught children had to fend for themselves in San Francisco (my mother, then fifteen, and her older sister had found work in a sewing factory), a lawyer was hired to fight for Grandmother's release from the detention barracks. A letter addressed to her on Angel Island from her attorney, C. M. Fickert, dated 24 March,1924, reads: "Everything I can legitimately do will be done on your behalf. As you say, it seems most inhuman for you to be separated from your children who need your care. I am sorry that the immigration officers will not see the human side of your case."
>
> Times were tough for Chinese immigrants in 1924. . . .
>
> The year my grandmother was detained on Angel Island, a law had just taken effect that forbade all aliens ineligible for citizenship

from landing in America. This constituted a virtual ban on the immigration of all Chinese, including Chinese wives of US citizens.

Waiting month after month in the bleak barracks, Grandmother heard many heart-rending stories from women awaiting deportation. They spoke of the suicides of several despondent women who hanged themselves in the shower stalls. Grandmother could see the calligraphy carved on the walls by other detained immigrants, eloquent poems expressing homesickness, sorrow, and a sense of injustice.

Meanwhile, Fickert was sending telegrams to Washington (a total of ten the bill stated) and building up a case for the circuit court. Mrs. Lee, after all, was the wife of a citizen who was a respected San Francisco merchant, and her children were American citizens. He also consulted a medical authority to see about a cure for liver fluke.

My mother took the ferry from San Francisco twice a week to visit Grandmother and take her Chinese dishes such as salted eggs and steamed pork because Grandmother could not eat the beef stew served in the mess hall. Mother and daughter could not help crying frequently during their short visits in the administration building. They were under the close watch of both a guard and an interpreter.

After fifteen months the case was finally won. Grandmother was easily cured of filariasis and allowed—with nine months probation—to join her children in San Francisco. The legal fees amounted to $782.50, a fortune in those days.

In 1927 Dr. Frederick Lam in Hawaii, moved by the plight of Chinese families deported from the islands because of the liver fluke disease, worked to convince federal health officials that the disease was non-communicable. He used the case of Mrs. Lee Yoke Suey, my grandmother, as a precedent for allowing an immigrant to land with such an ailment and thus succeeded in breaking down a major barrier to Asian immigration.

My most vivid memory of Grandmother Lee is when she was in her seventies and studying for citizenship. She had asked me to test her on the three branches of government and how to pronounce them correctly. I was a sophomore in high school and had entered the "What American Democracy Means to Me" speech contest of the Chinese American Citizens Alliance. I looked directly at my grandmother in the audience. She didn't smile, and afterwards, didn't comment on my patriotic words. She had never told me about being on Angel Island or about her friends losing their citizenship. It wasn't in the textbooks either. I may have thought she wanted to be a citizen because her sons and sons-in-law had fought for this country, and we

> lived in a land of freedom and opportunity, but my guess now is that she wanted to avoid any possible confrontation—even at her age—with immigration authorities. The bad laws had been repealed, but she wasn't taking any chances.[1]

CONNECTIONS

Yu's grandmother was not a penniless immigrant nor was she ignorant of American ways. Most immigrants would not have known how to find a lawyer or had the money to pay one. What does her story suggest about the vulnerability of immigrants—particularly immigrants who have been defined as one of *them*?

Why do you think Connie Young Yu's grandmother decided to become a citizen when she was in her seventies? How does Yu explain that decision? What does her explanation suggest about the dangers of being seen as outside a nation's universe of obligation—the circle of individuals and groups toward whom it has obligations, to whom the rules of society apply, and whose injuries call for amends?

The Chinese were the first immigrants to be excluded from the United States. Those already in the nation experienced prejudice and discrimination. Connie Young Yu explains one of the consequences of lying beyond a nation's universe of obligation.

> In Asian America there are two kinds of history. The first is what is written about us in various old volumes on immigrants and echoed in textbooks, and the second is our own oral history, what we learn in the family chain of generations. We are writing this oral history ourselves. But as we research the factual background of our story, we face the dilemma of finding sources. Worse than burning the books is not being included in the record at all, and in American history—traditionally viewed from the white male perspective—minority women have been virtually ignored.[2]

Why do you think she views being excluded as "worse than burning the books"? How do those who are left out find their place in the history books?

---

1. "The World of Our Grandmothers" by Connie Young Yu in *Making Waves* ed. by Asian Women United of California © 1989 by Asian Women United of California. Reprinted by permission of Beacon Press, Boston, pp. 39–41.
2. Ibid., p. 41.

Reading 9

The United State today is in many ways a very different country than it was in Harry Laughlin's day. Although myths and misinformation about the "other" continue, state and federal laws now ban most forms of discrimination and outlaw segregation. The change is also reflected in the nation's immigration laws. In 1965 Congress replaced the old quota system established by the National Origins Act of 1924. The old law was racist. It favored immigrants from Western Europe over those from other parts of the world. It literally cut off all immigration from Asia and Africa. The new law ended that discrimination by establishing a system that gives preferences to refugees from all parts of the world, people with relatives in the United States, and workers with needed skills.

The results of the 2000 Census reveal how the new law has altered the nation in small ways and large. Today only 16 percent of the nation's foreign born are from Europe. A little over half (51 percent) come from Latin America, 27 percent from Asia, 16 percent from Europe, and 6 percent from other areas of the world. Unlike earlier arrivals, the newcomers have not settled in cities or on farms but in the suburbs. Reporter Rick Hampson notes:

> A hundred years ago, immigrants from India might have moved onto six blocks on New York's Lower East Side. Now they move into six neighborhoods in central New Jersey.
>
> Instead of walking among pushcarts on Orchard Street, immigrants drive Toyotas to mini-malls filled with stores where their language is spoken.
>
> Many residents of Los Angeles' Koreatown are now Hispanic. Korean immigrants fan across the L.A. basin and form satellite settlements that together constitute the largest Korean community outside Korea. . . .
>
> In Garden City, a southwest Kansas community of about 30,000, City Hall has signs in English, Spanish and Vietnamese. Immigrants from Mexico and Southeast Asia have been attracted by meatpacking houses, which offer work few Americans want to do.[1]

The newcomers are changing many parts of American culture, including attitudes toward race. Cindy Rodriguez, a reporter for the *Boston Globe*, describes those changes in an article that focuses on Lawrence, Massachusetts:

> They call this Platano City, a place where bins at the corner bodegas overflow with platanos, the green plantains that Latinos from

the Caribbean smash into discs and then fry.

Latinos dominate here, making up an estimate 60 percent of the population, the epicenter of a Latino boom north of Boston.

Throughout the country, communities such as this have made Latinos the nation's fastest-growing minority changing the culture and flavor of urban centers, and in a subtle way, altering the way Americans look at race.

Regardless of their ancestral makeup, whether they have strong African features or more Spanish blood, Latinos don't view themselves as strictly black or white, largely freeing them from the us vs. them mentality that colors U.S. race relations.

"People in America get caught up in race," said Felix Coto, 17, a dark-skinned Latino, walking along Broadway with his girlfriend, Ramona Fernandez, who is light-skinned.

"I don't see him as black," Fernandez said. "He is Dominican, just like me."

[At the end of 2000], the US Census Bureau announced the nation's population stood at 281 million—6 million higher than anticipated.

One of the driving forces behind the growth is the influx of immigrants from Latin America, which helped give Texas and Arizona two extra representatives each in the 435-member US House of Representatives at the expense of the slower-growing Northeast.

"Latinos will play an important role in changing the way America defines race," said Clara E. Rodriguez, a Fordham University professor and author of *Changing Race: Latinos, the Census and the History of Ethnicity in the United States.*

"Race is a social construct, and because of that it will change over time," she said.

Rodriguez said when large numbers of Latinos rejected race categories on the census and checked off "other race" in the past, many people thought they were confused. But it has become clear, she explained, that Latinos see themselves as stretching across racial lines, fitting in two or even three categories.

This isn't a case of cultural pride, or about ethnicity trumping color, Latino scholars say. It's about a mindset of racial fluidity that contrasts with America's legacy of slavery and its painful aftermath. Although racism—against indigenous people and those who are dark-skinned—is evident in Latin America, especially in disparate poverty rates, Latin America did not have overtly racist laws. There were no Jim Crow voting laws that disenfranchised minorities, no segregated schools, no separate water fountains.

Latinos tend to look at skin color not as fixed markers of race, but as a continuum that shows the melange of ethnic groups that resulted in an endless array of hues.

"It's not viewed as starkly," said William Javier Nelson a Dominican who teaches sociology at Shaw University in Raleigh, N.C., a historically black college. "It's not until Latinos came to the US that they're confronted with the black-white dichotomy."

Latino immigrants say that it wasn't until they arrived in the United States that they began to face the polarizing aspects of race.

Before Gustavo Reyes, 32, emigrated from the Dominican Republic six years ago, he viewed himself as Dominican and Latino. But once he stepped off the plane at Logan International Airport, he was suddenly viewed by others as black.

Similarly, Regia Gonzalez, a Cuban who arrived in 1971, also saw herself as Cuban and Latino. But once in the United States, she was labeled white.

Neither Reyes, whose cocoa complexion shows his African ancestry, nor Gonzalez, whose great-great-grandparents hail from Spain, accepts the US-given racial designation.

In their eyes, they are Latino.

"I don't like the terms 'black Hispanic' and 'white Hispanic.' What is that?" said Reyes, a DJ for La Mega, a Spanish-language AM radio station. "I don't know too many Latinos who are pure white or pure black."

Latinos viewed themselves as multiracial long before Tiger Woods popularized the concept in America. They don't adhere to America's "one drop rule" which, throughout history, would categorize anyone with a smidgeon of African blood as a black person.

Americans use the mutually exclusive terms black and white, which have a polarizing effect, scholars say. They say it's interesting to note that "brown" has become the figurative word to identify Latinos, who are wedged in the middle. Though Latinos use the terms "negro" and "blanco" as well, they are more likely to refer to skin tone, not political outlooks. In the same way that they use "moreno" for a person with brown skin color, and "trigueno" for a person with tawny skin.

Latinos use the term "la raza"—literally "the race"—to refer to the wide spectrum of people who comprise Latinos, from Peruvians, who have more Andean blood, to Dominicans who have more African blood, to Chileans, who have more Spanish blood.

"In Cuba, there was no difference between a black Cuban, and a white Cuban and a Chinese Cuban," said Gonzalez, 52 of Roslindale, noting the large Chinese immigrant population. "We were all Cuban. My best friend was a so-called black Cuban. But my parents would never have told me, 'Don't have a friend who has dark skin.' It's not like in this country."

Some people would argue that Gonzalez is overlooking the racism that exists in Cuba and throughout Afro-Hispano countries, but dark-skinned Latinos are the first to say that they are confronted by race in the United States, more so than back home. . . .

[Nelson] said that in cities like Raleigh, which has a small Latino population made up mostly of Mexicans, he gets labeled as black. He rejects the term because it doesn't accurately reflect his history, his culture.

"It's one thing to say you are part of the African diaspora and another to say you are black," Nelson said.

At times, African Americans tell him he is rejecting his blackness, but he doesn't see it that way. He said he thinks of American blacks and Africans as his "long-lost cousins," and that he embraces his African heritage, but says it is just one part of him.

For light-skinned Latinos, there is a different reaction to their being placed in a racial category. Many reject being called white because don't like being associated with "the oppressor"—the Spanish who conquered and colonized much of Latin America.

"If I'm categorized as a 'white Hispanic,' then they are saying I am a Spaniard. And I am not," said Will Morales, 30, of Roslindale, who is a beige-skinned Puerto Rican. "I don't view myself as white. I relate more as a person of color."

He prefers the term Latino, which he says, "transcends the color piece."

But the majority of Latinos in the United States, well over 60 percent, don't have African ancestry, but rather so-called "Indian" ancestry. That includes Mexicans, the largest Latino group in the United States, and Central Americans, who have been arriving in large waves since the 1980s. Because of the demand for workers in the hospitality industry, large influxes have been arriving in New England.

Julio Cesar Aragon, a Mexican who arrived from Chihuahua more than 20 years ago, doesn't know how to classify himself in the United States. His national origin is Mexican, his ethnicity is Latino, but his race is "Indian."

"I am a descendant of the Tarhaumaras," said Aragon, 37, the president of the Mexican Association of Rhode Island. The racial box "Native American" doesn't apply to him, he says, because that refers to indigenous people of North America, such as Cherokee and Onondaga.

But he, too, doesn't want to be placed in one racial category. "What white or black people think of us doesn't matter. . . . I know what I am," Aragon said.[2]

CONNECTIONS

Harry Laughlin told a congressional committee, "If the American Nation . . . decides that we have national ideals worth saving, not only in national tradition and individual quality, but also racial ingredients, the Nation must exercise stricter control over immigration." The Immigration Act of 1924 was the result of that view of race and citizenship. It was based on a belief that "race" is a scientific construct. In what ways did the Immigration Act of 1965 challenge that view? How have the new immigrants challenged it?

Who in a society determines which differences matter? Where do we get our ideas about "race"? How do we learn what is "normal"? How do we decide who is beautiful? What part does family play? What is the role of the media? To what extent do media images shape standards of beauty? To what extent are those images a reflection of the views of society?

Interview an immigrant who has come to United States since 1965. What does his or her story add to your understanding of immigration today? What questions does it raise? Share your interview with your classmates and discuss similarities among the people you interviewed. How do you account for differences?

---

1. "1990s Boom Reminiscent of 1890s" by Rick Hampson. *USA Today*, May 24, 2001.
2. "Latinos Give US New View of Race" by Cindy Rodriguez. *Boston Globe*, January 2, 2001.

8. The Nazi Connection

Eugenics is not a panacea that will cure human ills, it is rather a dangerous sword that may turn its edge against those who rely on its strength.

Franz Boas

When Francis Galton founded the eugenics movement, he hoped that his new branch of scientific inquiry would someday become an international movement. By the early 1900s his dream was becoming a reality. There were now eugenicists in nations around the world. The movement was particularly popular in the United States and Germany.

Many Americans were intrigued at the notion of a "panacea that will cure human ills." They found it all too easy to believe that scientists and politicians could work together to solve social problems by mandating racial segregation, sterilizing the "feebleminded," and closing the nation's borders to "inferior hordes of degenerate peoples." After all, they reasoned, such laws were supported by research and endorsed by scholars at leading universities. Critics of eugenics were mostly ignored, as the nation led the world in eugenics research.

Although Germans were also flattered at the idea of belonging to a "superior race," few expressed interest in the movement until after World War I. Bitter and angry at the nation's losses, many looked for someone to blame. Some turned against "the Jews" and other "racial enemies." Others directed their anger toward the "useless eaters" who stayed at home while the nation's finest young men were murdered on the battlefields. In their efforts to protect the "race" by "breeding the best with the best," these Germans found inspiration and encouragement in the eugenics movement. By the 1920s German and Americans eugenicists were working side by side on a variety of research projects.

Eugenics also influenced the thinking of political leaders in both nations. Throughout the early 1900s eugenics had the support of American presidents and lawmakers. In Germany, it was central to the programs advocated by Adolf Hitler and his Nazi party. When Hitler came to power in 1933, he used eugenic principles to build a "racial state." Ironically, he applied those principles to German life at a time when scientific discoveries were undercutting both eugenics and racism. Jacob Landsman, an American critic of eugenics, summarized the new insights in the mid-1930s:

> It is not true that boiler washers, engine hostlers, miners, janitors, and garbage men, who have large families, are necessarily idiots and morons. . . . It is not true that celebrated individuals

> necessarily beget celebrated offspring . . . [or] that idiotic individuals necessarily beget idiotic children. . . . It is not true that, because the color of guinea pigs is transmissible in accordance with the Mendelian theory, therefore human mental traits must also be. . . . It is not true that, by any known scientific test, there is a Nordic race or that the so-called Nordic race is superior to any other race.[1]

Landsman might have added that it is also not true that sterilizing the "unfit" will end or even reduce social problems. Yet most people in the United States and Germany were unaware that Landsman and a number of other scientists no longer considered eugenics "scientific." Although eugenicists were eager to share their views and influence legislation and social policy, few other scientists were willing to speak out on the issues of the day. Their silence had real consequences.

This chapter raises important questions about the relationship between science and society at a time when Hitler was determined to annihilate Jews and other "racial enemies." In reflecting on that relationship in 1939, U.S. Vice President Henry Wallace asked: "Under what conditions will the scientist deny the truth and pervert his science to serve the slogans of tyranny? Under what conditions are great numbers of men willing to surrender all hope of individual freedom and become ciphers of the State? How can these conditions be prevented from occurring in our country?" Many of the readings in this chapter explore the ways scientists, political leaders, and ordinary citizens answered those questions in the 1930s and 1940s.

---

1 Quoted in *In the Name of Eugenics* by Daniel J. Kevles. Harvard University Press, 1985, 1995, p. 164.

Reading 1

The early 1900s were years of unrest throughout the world. Economic dislocations, global war, fears of an international Communist revolution, and by the early 1930s, a worldwide depression threatened stability everywhere. With uncertainty came doubts. Convinced that democracy had failed, some turned to communism. Others were attracted to fascism. Fascists insisted that democracy puts "selfish individual interests" before the needs of the nation. They placed their faith in a leader who stood above politics.

In 1922, Benito Mussolini established the world's first fascist government in Italy. It would later serve as a model for the one Adolf Hitler set up in Germany. In both nations, the word of the leader or führer was law. He was a dictator—a leader who was not dependent on a legislature, courts, or voters. According to Hitler, a *führer* or a *duce* (in Italian) is a leader "in whose name everything is done, who is said to be 'responsible' for all, but whose acts can nowhere be called into question," because "he is the genius or the hero conceived as the man of pure race."

Both Mussolini and Hitler maintained that only a few men are intelligent enough to rise in the world and that those men have an obligation to rule. In their view, decision making was too important to be left to the people. These ideas were attractive to a number of eugenicists. Throughout the 1920s, many of them traveled to Rome to meet with Mussolini. At one meeting in 1929, Charles Davenport, then president of the International Federation of Eugenic Organizations, and Eugen Fischer, a noted German eugenicist at the Kaiser Wilhelm Institute for Anthropology, Human Heredity and Eugenics, honored Mussolini. Davenport explained why they did so:

> The gravest concern of all eugenicists today is the preservation of human quality. It is a possibility! And in view of the tremendous importance for the future of every nation of this objective, no economic sacrifice can be too great. The sacrifices, however, would not be so very considerable. Here it is only possible to suggest how suitable measures in the sphere of property and income tax, and yet more certainly the inheritance tax can be brought to bear on maintaining families of talent in every social stratum. Such measures, however, should be fitted to the social position of the family, and favor those who have arrived at high position, and require to be so graded to the social rank attained that the best receive the greatest acknowledgement. Such suggestions may seem to sound an anti-social and

> anti-democratic note. It must, therefore, be borne in mind that each stratum in turn supplies its quota of those favored individuals who have attained social distinction, and the protection and advantages have to do with the family rather than with the individual—the family giving to the State children from amongst whom future leaders can be chosen. Thus every such attempt is in the truest sense of the word one which concerns "res publica"—in the highest sense democratic. Such administrative and legislative means are without doubt at hand, and can for each country be formulated by those forces in eugenics, in such a way that the legislators can make use of them.[1]

CONNECTIONS

Based on what you know about eugenics, why do you think Davenport views a "fall in the birth rate of the upper classes" as "catastrophic"?

Davenport describes himself and his colleagues as "men of science." How does he seem to view their role in society? In Chapter 7, physicist Leon Lederman was quoted as saying, "We [scientists] give you a powerful engine. You steer the ship." With what parts of that statement might Davenport agree? How does he define the role of a citizen? Who does he believe should "steer the ship"?

According to Davenport, which of his ideas sound "anti-social" and "anti-democratic"? How does he defend those ideas? What does his defense suggest about the way he views democracy? The way he regards the relationship between science and society? To what extent does his defense explain why he was so eager to win over a dictator like Mussolini?

Just five years before Charles Davenport's meeting with Mussolini, he and other eugenicists persuaded Congress to strictly limit the number of Italians who could settle in the United States. Why then would Davenport single out Mussolini for praise as a "statesman"?

Draw a diagram showing how power is divided in a democracy. Who holds the power to make laws? Enforce laws? The power to interpret the law? What role do ordinary citizens play? Draw a similar diagram showing the division of power in a fascist state. What role do ordinary citizens play? What part do leaders play? How well does either diagram square with reality?

---

1 Quoted in *The Legacy of Malthus: The Social Costs of the New Scientific Racism* by Allan Chase. Alfred A. Knopf, 1977, pp. 346-347.

Reading 2

Eugenicists held their first international conference in London in 1912. It was an appropriate place for a meeting devoted to "race improvement." After all, Britain was the home of Francis Galton and the place where the eugenics movement began. Yet it was the Americans, not the British, who took center stage at the conference.

Delegates from other nations were impressed by the gains the United States had made in "protecting the race." Between 1907 and 1912, eight states had passed laws authorizing or requiring the sterilization of "certain classes of defectives and degenerates" and several others were considering similar legislation. American eugenicists also boasted of financial backing from private foundations and public agencies. Not surprisingly, in the years that followed the convention, Americans took over the leadership of the International Congress of Eugenics. The first president was an Englishman—Leonard Darwin, the son of Charles Darwin and a cousin of Francis Galton. The group's second and third presidents were Americans—Henry F. Osborn and Charles Davenport.

Even before the meeting, many European eugenicists were closely following events in the United States. The Germans were particularly interested in the American experience. In Germany, eugenics was known as "racial hygiene." Alfred Ploetz, the founder of the movement, was a physician who believed that governments were allowing "the least fit" in society to survive at the expense of the "fittest." To address the problem, he advocated a new kind of hygiene—one that promoted the health not only of the individual but also of the "race."

Throughout the early 1900s, Ploetz and his followers organized meetings dedicated to "race improvement," published journals that promoted eugenics, and built formal and informal relationships with like-minded scholars at home and abroad. In 1905, they founded the Society for Racial Hygiene. A few years after the first international conference in London, the Berlin branch of the society distributed a brochure lauding "the dedication with which Americans sponsor research in the field of racial hygiene and with which they translate theoretical knowledge into practice." The document also praised the nation's "fantastic" control of immigration through restrictive laws and applauded the American states that had statutes designed to keep "inferior families" from having children. The brochure ended with a question: "Can we have any doubts that the Americans will reach their aim—the stabilization and improvement of the strength of the people?"[1] The unspoken question was: Would Germans do the same?

Géza von Hoffman, an Austrian diplomat based in California in the early 1900s, provided the society with much of its information about the American eugenics movement. During his stay in the United States, he wrote numerous articles and, in 1913, a book on the topic. He was not the only German to look to the United States for lessons on applying eugenics to public policy. In the early 1900s, German medical authorities gathered information about state laws that banned marriages if one partner was alcoholic, "feebleminded," insane, or suffered from such diseases as tuberculosis or syphilis. The Reich Health Office even kept a special file on such laws. As more states passed "eugenic laws," the file grew. So did the number of Germans who visited the United States to observe "eugenics in action" and the number of books by American eugenicists that were translated into the German language.

California eugenicist Paul Popenoe explains a pedigree chart, 1930.

By the time World War I began in 1914, Americans had established their leadership in the eugenics movement and laid the foundation for international cooperation. After the war ended, eugenicists on both sides of the Atlantic Ocean were eager to reestablish old ties and forge new links. The war had convinced many Germans of the importance of "racial hygiene." They feared that the nation had lost its best young men on the battlefield while the "unfit" were protected at home. In Germany, medical care was under government control. Therefore the taxpayers provided the money for the care of the physically and mentally disabled. The economic crises of the 1920s and 1930s in Germany added to many people's sense of outrage. These Germans insisted that the cost of supporting the "unfit" was a growing burden on the entire nation.

American eugenicists encouraged Germany's interest in finding "biological" solutions to the nation's problems. Charles Davenport led the effort by working to reintegrate the Germans into the international eugenics movement despite resistance from many of Germany's opponents in World War I. At the same time, he promoted joint research with his German counterparts on a variety of projects.

CONNECTIONS

The word *hygiene* refers to practices and conditions that promote health. What then is *racial hygiene*? What words or phrases come to mind when you think of "good hygiene"? "Poor hygiene"? How do you think having a physician like Ploetz link eugenic ideas to health, cleanliness, and physical well-being shaped public opinion about the disabled, the mentally ill, and other "misfits"?

Scholarly organizations play an important part in shaping public opinion. How did such groups encourage the spread of eugenic ideas?

How do you think the labeling of groups as "inferior" or a "burden" on society may have shaped the way individuals saw themselves as "others"? What effect might that kind of labeling have on the way Germans defined their "universe of obligation"?

How were the efforts of the German and American eugenicists to "protect the race" similar? What differences seem most striking?

---

1. *The Nazi Connection* by Stefan Kuhl. Oxford University Press, 1994, pp. 15-16.

"Thinking Biologically"

Reading 3

In 1921, Fritz Lenz, Eugen Fischer, and Erwin Baur published a two-volume work entitled *Outline of Human Genetics and Racial Hygiene*. Reviewers hailed the work as a "masterpiece" in the best traditions of German scholarship. Revised and updated every few years, the work shaped medical thinking in Germany and provided scientific legitimacy for Adolf Hitler's National Socialist or Nazi party. Indeed the publisher sent Hitler a copy of the 1923 edition. He read it during the year he spent in prison for an attempted overthrow of the German government. Later, in reviewing *Mein Kampf*, Hitler's own account of his political beliefs about German racial superiority and his dreams of building a new Germany empire, Fritz Lenz noted with pride that Hitler had borrowed many of his own ideas.

From top: Eugen Fischer, Erwin Baur, and Fritz Lenz.

Throughout their work, the three authors acknowledge American leadership in the eugenics movement. They repeatedly cite research by such American scholars as Henry Goddard (Chapter 3), Charles Davenport (Chapter 3), Carl Brigham (Chapter 5), and Lewis Terman (Chapter 5). Lenz, in particular, insisted that there were no differences between the positions taken by American and German eugenicists. Both were "accustomed to thinking biologically."[1] Although Germany lagged behind in the application of eugenics to public policy, he was confident that as eugenic education proceeded in Germany, eugenic laws would follow.

Throughout the 1920s and 1930s, Lenz, Baur, Fischer, and other German eugenicists worked closely with their American counterparts, especially Charles Davenport and Harry Laughlin at the Eugenics Record Office in Cold Spring Harbor, New York, and Paul Popenoe, a leader in the American Eugenics Society in California. A topic of mutual interest was "race crossing" or miscegenation. In 1929, Davenport invited Eugen Fischer to speak on the subject at the Rome meeting of the International Federation of Eugenic Organizations (IFEO).

Fischer had been active in the German eugenics movement since the early 1900s. Trained as an anthropologist at Freiburg University, he led a research team to what was then the German colony of Southwest Africa, now Namibia. He arrived in 1909, shortly after German soldiers had murdered about 75 percent of the Herero people—children, women, and men. Fischer had little interest in this genocide. He focused instead on the offspring of marriages between Dutch men and Herero women—the so-called "Rehoboth Bastards" despite the fact that their parents were legally married. Fischer measured their heads, took blood samples, and then compared the results to similar measurements taken from the surviving Hereros. Claiming that children of so-called "mixed marriages" were of "lesser racial quality," he insisted that their intellectual achievements were directly related to the amount of "European blood" in their veins. In 1913, he concluded:

> Without exception, every European people that has accepted blood from inferior races—and the fact that the Negroes, Hottentots and many others are inferior can be denied only by dreamers—has suffered an intellectual and cultural decline as a result of the acceptance of inferior elements.[2]

Fischer's research led to his appointment as the director of a department in the newly established Kaiser Wilhelm Institute for Anthropology, Eugenics, and Human Heredity just after the war. It had the backing of the Rockefeller Foundation of New York, which supported a number of other eugenics research institutes in Germany in the 1920s. At the official opening of the Kaiser Wilhelm Institute, Fischer and his colleagues invited Charles Davenport to speak. Honored by the invitation, Davenport used the occasion to promote further research on the eugenic consequences of miscegenation. In 1928, researchers at Davenport's Eugenics Record Office in New York and the Kaiser Wilhelm Institute in Berlin prepared a questionnaire for distribution to one thousand physicians, missionaries, and diplomats around the world. They hoped to gather global data on the effects of "race mixing." Davenport and Fischer also formed a Committee on Race Crossing within the IFEO. Fritz Lenz chaired the group and urged further research on intermarriages with Jews.

Despite their scholarly achievements, German eugenicists in the 1920s encountered strong religious and social opposition whenever they tried to translate their research into public policy. After the U.S. Congress passed the 1924 National Origins Act (Chapter 7), a Bavarian health inspector wistfully noted, "German racial hygienists should learn from the United States how to restrict the influx of Jews and eastern and southern Europeans."[3] The law also won praise from Adolf Hitler who praised the act for its exclusion of "undesirables" on the basis of hereditary illness and race.

When Hitler came to power in 1933, Lenz hailed him as the first politician "of truly great import, who has taken racial hygiene as a serious element of state policy."[4] He and other German eugenicists saw Hitler's rise as an opportunity to make their nation "the first in world history" to apply "the principles of race, genetics, and selection to practical politics." Although Lenz and others initially expressed some reservations about Hitler's antisemitism, they actively supported the new regime. They wrote essays and books in defense of Nazi policies, took an active role in designing eugenic laws and decrees, and then helped the Nazis implement those measures. In 1938, Theodor Mollison, the director of the Anthropological Institute in Munich, defended their support for Hitler in a letter to Franz Boas, a critic of the eugenics movement in general and the Nazis in particular:

> If you think that we scientists do not agree with the cry, "Heil Hitler," then you are very much mistaken. If you would take a look at today's Germany, you would see that progress is being made in this Third Reich, progress that never would have come to pass under the previous regime, habituated as it was to idleness and feeding the unemployed instead of giving them work. The claim that scientific thought is not free in Germany is absurd. . . . I assure you that we German scientists know well the things for which we may thank Adolf Hitler, not the least of which is the cleansing of our people from foreign racial elements, whose manner of thinking is not our own. With the exception of those few individuals with ties to Jewish or Masonic groups, we scientists support wholeheartedly the salute "Heil Hitler."[5]

CONNECTIONS

In 1930, Carl Brigham wrote an article in which he retracted many of the conclusions he had reached in *A Study of American Intelligence*. "Comparative studies of various national and racial groups may not be made with existing tests," he now argued. He went on to state that one of the most pretentious of these comparative racial studies—the writer's own—was without foundation."[6] (page 174) Why do you think Lenz, Baur, and Fischer ignored Brigham's retraction when they revised their book in the early 1930s? What does their action suggest about the quality of their research?

In the early 1900s, the Germans committed genocide in Southwest Africa. What does the word *genocide* mean? Record your definition in your journal so that you can revise and expand it as you continue reading.

Fischer saw firsthand the effects of the genocide in Africa. Yet he made no mention of it in his research on a related topic—intermarriages between Dutch men and Herero women. What does his silence suggest about the way he approached his work as a scientist? About the way he defined his universe of obligation?

What opportunities did German eugenicists see in Hitler's rise to power? What advantages do you think Hitler may have seen in their support? For more information, consult Chapters 4 and 5 of *Facing History and Ourselves: Holocaust and Human Behavior.*

Davenport worked with Fischer and Lenz long before the Nazis took power in Germany. How do you think their collaboration might have changed once Hitler consolidated his power? What opportunities do such collaborations provide? What are the risks in such collaborations?

In the 1920s, many Germans looked back on their defeat in World War I and tried to explain it away. They came to believe that Jews had betrayed the nation. Initially, antisemitism was not a large part of the German eugenics movement. Now it became a cornerstone of German eugenics.[7] How do you account for the shift? To what extent was scientific opinion leading a social trend? To what extent was it following a social trend? What does Mollison's letter add to your understanding of the shift?

Franz Boas was a German-born anthropologist who was outspoken in his contempt for the Nazis. He was also a Jew. How do you think he responded to the letter from Mollison? To the idea that he and other Jews were "foreign racial elements" that ought to be cleansed from German society?

---

1. Quoted in *Racial Hygiene: Medicine under the Nazis* by Robert Proctor. Harvard University Press, 1988, p. 50.
2. Quoted in *The Value of the Human Being: Medicine in Germany 1918-1945* by Christian Pross and Gotz Aly. Arztekammer Berlin, 1999, p. 15.
3. *The Nazi Connection* by Stefan Kuhl. Oxford University Press, 1994, p. 26.
4. Quoted in *Racial Hygiene: Medicine under the Nazis* by Robert Proctor. Harvard University Press, 1988, p. 61.
5. Quoted by Robert Proctor, "From Anthropologie to Rassenkunde" in *Bones, Bodies, Behavior* edited by George W. Stocking, Jr.. University of Wisconsin, Press 1988, p. 166.
6. "Intelligence Tests of Immigrant Groups" by Carl Brigham. *Psychological Review* 37, 1930, p. 165.
7. "Eugenics Among the Social Sciences: Hereditarian Thought in Germany and the United States" by Robert Proctor in *The Estate of Social Knowledge* ed. by J. Brown and D. K. Van Keuren. John Hopkins University Press, 1991, p. 89.

Reading 4

Adolf Hitler believed that "the race question" is the key to world history and world culture. He insisted that society is based on the struggle of the "lower races" against the "higher races." Who were the "lower races"? To Hitler, the answer was clear: they were Eastern Europeans, Africans, "Gypsies," and Jews. These ideas about the superiority of the "Aryan" or "Nordic" race were not new. They were taught in German schools and universities long before Hitler came to power. Hitler was the first, however, to take German scientists and other scholars at their word. From the start, he declared that he would protect the purity of the "Aryan" race from its "racial enemies" by turning Germany into a "racial state." That decision affected virtually every institution in the country and eventually became part of the Nazis' rationale for the Holocaust—the mass murder of millions of Jews, "Gypsies," and other "inferior peoples." The timeline below details Hitler's efforts to build a "racial state"—step by step, law by law, decree by decree.

BUILDING A RACIAL STATE: A TIMELINE

1933

January: The Nazi party takes power in Germany. Adolf Hitler becomes chancellor.
February: Nazis "temporarily" suspend civil liberties. They were never restored.
March: The Nazis set up the first concentration camp at Dachau. The first inmates are 200 Communists.
April: The Nazis announce a one-day boycott of Jewish businesses. The Nazis enact the Civil Service Law, requiring proof of Aryan ancestry and political reliability to hold a government job.
July: The Nazis pass the "Law for the Prevention of Genetically Diseased Offspring," allowing for the compulsory sterilization for "eugenic reasons" of the "feebleminded," schizophrenics, alcoholics, and other carriers of supposedly single-gene traits.

1934

The government offers special loans to "racially sound" married men whose wives agree to give up jobs outside the home. For each child the government forgives 25 percent of the principal owed on the loan.

August: Hitler combines the positions of chancellor and president to become Fuhrer.

November: The "Law against Dangerous Career Criminals" permits the detention and castration of sex offenders and others guilty of "racial-biological" crimes.

1935

June: The "Law for the Alteration of the Law for the Prevention of Genetically Diseased Offspring" sanctions compulsory abortion, up to and including the sixth month of pregancy, for women categorized as "herditarily ill."

September: The "Law for the Protection of German Blood and German Honor" bars marriage and sexual relations between Aryans and Jews, "Gypsies," Africans, and their offspring.

The "Citizenship Law" distinguishes between citizens and "inhabitants." Jews and other non-Aryans are defined as "inhabitants" and deprived of citizenship rights.

October: The "Law for the Protection of the Hereditary Health of the German People" requires the registration and exclusuion of "alien" races and the "racially less valuable" from the "national community."

Before a marriage can take place, public health officials have to issue a "certificate of fitness to marry."

1936

March: German soldiers occupy the Rhineland, a buffer zone between Germany and France and Belgium established after World War I.

1938

January: The government withdraws the licenses of all Jewish physicians.

March: German troops annex Austria.

April: Jews are banned from almost every profession in Germany and

Austria. Jews are required to carry special papers identifying them as Jews.

November: On Kristallnacht, the night of the 10th-11th, Nazis gangs attack Jews throughout Germany and Austria, looting and then burning homes, synagogues, and businesses. They kill over 90 Jews and send over 30,000 others to concentration camps.

Jews are ordered to pay damages from the events of Kristallnacht.

Jews are barred from theaters, concerts, circuses, and other public places, including schools.

1939

March: Germany takes over Czechoslovakia

September: Germany invades Poland. World War II begins in Europe. Hitler secretly orders the systematic murder of the mentally and physically disabled in Germany and Austria.

December: Polish Jews are forced to relocate. They are also required to wear armbands or yellow stars.

1940

January: German physicians begin gassing mental patients, using carbon monoxide gas in fake showers in a psychiatric hospital near Berlin. The program is carried out under the code name T4 (the abbreviated address of the head of Hitler's "euthenasia program"). By September, over 70,000 were dead.

Spring: Approximately 30,000 people are killed at Hartheim, a mental hospital in Austria.

Nazis begin deporting German Jews to Poland.

Jews are forced into ghettoes.

June: The Nazis begin gassing Jews. The first 200 are from a mental institution.

Germany conquers much of Western Europe.

1941

German psychiatrists train the SS, the Nazis' elite troops, on mass murder techniques learned from experimentation on mental patients.

The Reich Interior Minister orders the killing of Jews in German mental hospitals. Roving bands of T4 commissions select those too ill to work as well as Jews and "Gypsies" in concentration camps and send them to gas chambers at psychiatric hospitals.
June: Germany invades the Soviet Union.

Jews throughout Europe are forced into ghettoes and internment camps.

Mobile killing units begin the systematic slaughter of Jews. In two days, one unit murders 33,771 Ukrainian Jews at Babi Yar—the largest single massacre of the Holocaust.

The first death camp at Chelmno in Poland begins operations.
December: After the Japanese bombing of Pearl Harbor, Hawaii, the United States enters World War II by declaring war on both Japan and its main ally, Germany.

1942

January: At the Wannsee Conference, Nazi officials turn over the "Final Solution"—their plan to kill all European Jews—to the bureaucracy.

Five death camps begin operation in Poland: Majdanek, Sobibor, Treblinka, Belzec, and Auschwitz-Birkenau.
December: The United States, Britain, and the Soviet Union acknowledge that Germans are systematically murdering the Jews of Europe.

1944

March: Hitler's troops occupy Hungary.
June: The Germans deport 12,000 Hungarian Jews a day to Auschwitz.

1945

January: As the Soviet army pushes east, the Nazis evacuate the death camps.
May: World War II ends in Europe with Hitler's defeat. Hitler's racial state is dismantled. About one-third of all Europe's Jews are dead and most of the survivors are homeless.

CONNECTIONS

Throughout the 1930s, Hitler advanced his plans to turn Germany into a racial state. When an action against an individual, group, or even a nation resulted in opposition, he quickly backed down. If he encountered little or no opposition, he was a little bolder the next time. Yet after Hitler's defeat, many people expressed surprise that he did exactly what he had promised to do. How do you account for their surprise? Why do you think they didn't try to stop him during his first years in power?

Many historians have noted that by the time many people were aware of the danger the Nazis posed, they were isolated and alone. What events on the timeline support that view? Notice the names given to the various laws included in the timeline. What do the titles reveal? What do they conceal? How might these laws be used to turn neighbor against neighbor?

Which laws listed would be particularly attractive to eugenicists in other countries? At what point do you think many of them might feel uneasy about their support for Hitler's policies? Record your ideas in your journal and review them as you examine the next few readings.

Reading 5

Just six months after Adolf Hitler took office, Germany enacted its first eugenic measure—the "Law for the Prevention of Genetically Diseased Offspring." The *Eugenical News,* which was published by the Eugenics Record Office, proudly printed a translation of the law. It states in part:

Paragraph 1.

(1) Whoever is afflicted with a hereditary disease can be sterilized by a surgical operation, if—according to the experience of medical science—there is a great probability that his descendants will suffer from serious bodily or mental defects.

(2) Hereditary diseases under this law are 1. Hereditary feeblemindedness, 2. Schizophrenia, 3. Manic-depressive insanity, 4. Hereditary epilepsy, 5. Huntington's Chorea, 6. Hereditary blindness, 7. Hereditary deafness, 8. Serious hereditary bodily deformities.

(3) Furthermore those suffering from Alcoholism can be sterilized.

Paragraph 2.

(1) Petition can be made by the subject to be sterilized. If this individual is incompetent, mentally deficient or has not yet completed his eighteenth year, a legal representative has the right to make application; the consent of the court of guardianship is required. In cases of limited competency the petition has to be approved by the legal representative. If the subject is of age and in charge of a caretaker, his consent is required.

(2) The petition is to be accompanied by a certificate from a physician recognized in the German Reich, testifying that the person nominated for sterilization has been enlightened on the nature and consequence of sterilization.

(3) The petition can be withdrawn.

Paragraph 3.

Sterilization can be requested by (1) the public health physician, (2) the superintendent, for the inmates of a hospital, custodial institution or a penitentiary.

Paragraph 4.

Petition is to be in writing or recorded with the District

Eugenical Court. The facts upon which the petition is made should be supported by a medical certificate or confirmed in some other way. The district court has to notify the public health physician.

Paragraph 5.

Decision rests with the Eugenical Court of the district to which the person nominated for sterilization belongs.

Paragraph 6.

(1) The Eugenical Court is to be part of a Tribunal. It consists of a judge, acting as chairman, a public health physician and another physician approved by the German Reich and particularly versed in Eugenics. An alternate is to be appointed for each member.

(2) As chairman must be excluded: one who has decided upon a petition from the court of guardianship according to Paragraph 2, item 1. If the public health physician has made the petition, he is excluded from the decision.

Paragraph 7.

(1) The proceedings of the Eugenical Court are not public.

(2) The Eugenical Court has to make the necessary investigations. It can hear witnesses and experts and order the personal appearance as well as a medical examination of the person to be sterilized, who can be summoned in case of unexcused absence. . . . Physicians who have been questioned as witnesses or experts are obliged to testify, regardless of medical ethics. Legal authorities as well as institutions have to give information to the Eugenical Court upon request.

Paragraph 8.

The court has to decide according to its free conviction, after considering the entire results of the procedure and testimony. The decision is based upon a majority of votes after verbal consultation. The court decision should be stated in writing and signed by the members acting as judges. The reasons for ordering or suspending sterilization must be indicated. . . .

Paragraph 9.

Persons designated in Paragraph 8, sentence 7, can take an appeal from the decision within a peremptory term of one month from the date of serving such notice. This appeal has a postponing effect.

The Supreme Eugenical Court decides upon this complaint. . . .

Paragraph 10.

(1) The Supreme Eugenical Court is part of the Supreme Court of the country and comprises its district. It consists of one member of the Supreme Court, one public health physician and one additional physician, approved by the German Reich, who is especially versed in Eugenics. . . . The judgment of the Supreme Eugenical Court is final.

Paragraph 11.

(1) The surgical operation necessary for sterilization should be performed only at a hospital and by a physician approved by the German Reich. This surgeon can perform the operation only when the order for sterilization has been made final. . . .

(2) The surgeon performing the operation has to submit a written report on the sterilization with a statement regarding the applied technique to the physician in charge.

Paragraph 12.

(1) When the court has finally decided upon the sterilization, the operation has to be performed even against the will of the subject to be sterilized, insofar as he has not made the petition alone. The public health physician has to attend to the necessary measures with the police authorities. . . .

(2) When circumstances arise requiring another trial of the case, the Eugenical Court has to resume the proceedings and temporarily suspend the sterilization. If this appeal has been rejected, resumption of proceeding is admissible only if new facts that have come to light justify the sterilization.

Paragraph 13.

(1) The costs of the court proceeding should be covered by the State funds.

(2) The cost of the surgical operation should be covered by the sick fund in the case of persons insured, and by the charity organization in the case of needy persons. In other cases the costs, up to the minimum doctors' fee and the average hospital fee of public hospitals should be paid by the State funds, beyond that by the sterilized individual.

Paragraph 14.

A sterilization not carried out according to the rules of this law [is] only permissible if performed by a skilled physician and for the avoidance of a serious danger to the life or health of the person on whom and with whose consent the operation has been performed.

Paragraph 15.

(1) Persons involved in the procedure or in the performance of the surgical operation are pledged to secrecy.

(2) Whoever acts against this ethical rule of silence shall be punished with imprisonment up to one year or fined.[1]

In a report on the new law funded by the Carnegie Foundations, the American Neurological Association noted: "It is fair to state that the Sterilization Act is not a product of Hitler's regime in that its main tenets were proposed and considered several years earlier before the Nazi regime took possession of Germany. There is no doubt that the Act conforms closely with present knowledge of medical eugenics."[2] Harry Laughlin of the Eugenics Record Office agreed. He praised the law in the *Eugenical News*:

> Doubtless the legislative and court history of the experimental sterilization laws in the 27 states of the American union provided the experience which Germany used in writing her new sterilization statute. To one versed in the history of eugenical sterilization in America, the text of the German statute reads almost like the "American model sterilization law." [3]

Laughlin and others believed that the German law was an improvement on American sterilization laws. In the United States, sterilization laws varied from state to state and enforcement was often inconsistent. The German measure, on the other hand, applied to the entire nation and promised to be uniformly enforced.

Before long, American eugencists were traveling to Germany to observe "eugenics in action." They visited "eugenic courts" and met with Nazi leaders as well as scholars and scientists. After his visit, Frederick Osborn, then secretary of the American Eugenics Society, hailed "recent developments in Germany" as "perhaps the most important experiment which has ever been tried."

Just a few months after the new law went into effect, Hitler called for the sterilization of "dangerous habitual criminals." Under cover of that law, the government sterilized individuals who had no physical or mental disability. These children, women, and men were targeted simply because they were "Gypsies,"

A 1934 exhibition in Pasadena, CA describes the role of eugenics in the "New Germany."

Germans of African descent, or Jews. For example, in 1937, the Nazis used the law to secretly sterilize all "German colored children." They were the offspring of German women and the African soldiers who occupied Germany after World War I.

By 1937, the Nazis had sterilized nearly 225,000 individuals—about 10 times the number sterilized in the United States over a 30-year period—partly because Nazi journals openly advised the "eugenical courts" not to be "over scrupulous" in their decisions.[4] They argued that it was better to make mistakes than jeopardize the "future" of the German people. So thousands of schizophrenics were sterilized, even though the classification of schizophrenia as a "hereditary disorder" was "no more than a working hypothesis," according to Hans Luxemberger, Germany's leading geneticist. He supported continued sterilization on the grounds that it might be too late when "final proof was established."[5] Despite signs that the Germans were sterilizing individuals with no "hereditary defects," American eugenicists remained convinced that Germany's sterilization law would never become an "instrument of tyranny."

CONNECTIONS

A euphemism is an inoffensive term used in place of a more explicit one. In Nazi Germany, euphemisms were used to disguise events, dehumanize "racial enemies," and diffuse responsibility for specific actions. Thus the Nazis did not speak of throwing their enemies into jail but of taking them into "protective custody." To what extent is the title of the new law a euphemism? What does it disguise or conceal? How does it regard the targets of the law? How does it diffuse responsibility for sterilization?

What is the role of a physician in the process outlined in the statute? Whom does the physician serve—the patient or the State? Why do you think the procedures of the eugenical courts were to be kept "secret"? Whose rights does a "secret proceeding" protect? Whose rights may such a proceeding threaten?

After visiting a hospital that performed sterilizations, Gregor Ziemer, an American educator, asked his SS guide who decides which women are to be sterilized. He was told, "We have courts. It is all done very legally, rest assured. We have law and order."[6] What does it mean to act "under the cover of the law"? What purposes do laws serve in a society? Are they a way of keeping order? Ensuring justice? Protecting rights?

What was the purpose of the sterilization law? How did it seem to alter traditional relationships in German society?

Compare Germany's "Law for the Prevention of Genetically Diseased Offspring" with Harry Laughlin's model sterilization law (Chapter 6). What similarities do you notice in the objectives of the two laws and the ways they are to be enforced? Which differences are most striking? Why do you think the American Neurological Association insisted that Germany's sterilization law was not a product of Hitler's regime?

---

1. *Eugenical News,* September-October, 1933.
2. Quoted in *By Trust Betrayed* by Hugh Gregory Gallagher. Holt, 1990, p. 93.
3. "Eugenical Sterilization in Germany," *Eugenical News*, September-October, 1933.
4. Quoted in *Hitler's Justice* by Ingo Müller. Trans. by Deborah Lucas Schneider. Harvard University Press, 1991, p. 122.
5. Ibid.
6. *Education for Death* by Gregor Ziemer. Oxford University Press, 1941, p. 28.

Reading 6

On September 15, 1935, the Nazis took another step toward protecting "Aryan blood" from "contamination." This time, they moved against the nation's Jews and other "racial enemies." It was not the Nazis' first anti-Jewish measure. They proclaimed 42 such laws in 1933 and 19 more in 1934. The new laws, which Hitler announced at a party rally in Nuremberg, provided the rationale for the earlier legislation. The first of these laws defined citizenship:

Article I
1. An inhabitant of the State is a person who belongs to the protective union of the German Reich, and who therefore has particular obligations towards the Reich.
2. The status of inhabitant is acquired in accordance with the provisions of the Reich and State Law of Citizenship.

Article 2
1. A citizen of the Reich is that inhabitant only who is of German or kindred blood and who, through his conduct, shows that he is both desirous and fit to serve the German people and Reich faithfully.
2. The right to citizenship is acquired by the granting of Reich citizenship papers.
3. Only the citizen of the Reich enjoys full political rights in accordance with the provision of the laws.

Article 3
The Reich Minister of the Interior in conjunction with the Deputy of the Führer will issue the necessary legal and administrative decrees for carrying out and supplementing this law.[1]

The second statute was the "Law for the Protection of German Blood and German Honor."

Section 1
1. Marriages between Jews and citizens of German or kindred blood are forbidden. Marriages concluded in defiance of this law are void, even if, for the purpose of evading this law, they were concluded abroad.
2. Proceedings for annulment may be initiated only by the Public Prosecutor.

Section 2
Sexual relations outside marriage between Jews and nationals of German or kindred blood are forbidden.

Section 3
Jews will not be permitted to employ female citizens of German or kindred blood as domestic servants.

Section 4
1. Jews are forbidden to display the Reich and national flag or the national colors.
2. On the other hand they are permitted to display the Jewish colors. The exercise of this right is protected by the State.

Section 5
1. A person who acts contrary to the prohibition of Section 1 will be punished with hard labor.
2. A person who acts contrary to the prohibition of Section 2 will be punished with imprisonment or with hard labor.
3. A person who acts contrary to the provisions of Sections 3 or 4 will be punished with imprisonment up to a year and with a fine, or with one of these penalties.

Section 6
The Reich Minister of the Interior in agreement with the Deputy Fuhrer and the Reich Minister of Justice will issue the legal and administrative regulations required for the enforcement and supplementing of this law. [2]

The laws raised an important question: Who is a Jew? In November, the Nazis defined a Jew as a person with two Jewish parents or three Jewish grandparents. Children with one Jewish parent were Jews if they practiced Judaism or married a Jew. A child of intermarriage who was not a Jew was a *Mischling*—a person of "mixed race." By isolating Jews from other Germans and forbidding mixing of races, the Nazis hoped that *Mischlings* would eventually disappear. The Nazis regarded these laws as public health measures. German medical journals often described miscegenation as a "public health hazard."

Regardless of their intent, the new laws and other antisemitic measures were successful. By the end of the year, at least a quarter of the Jews in Germany "had been deprived of their professional livelihood by boycott, decree, or local pressure," writes historian Martin Gilbert.

More than 10,000 public health and social workers had been driven out of their posts, 4,000 lawyers were without the right to practice, 2,000 doctors had been expelled from hospitals and clinics, 2,000 actors, singers, and musicians had been driven from their orchestras, clubs and cafes. A further 1,200 editors and journalists had been dismissed, as had 800 university professors and lecturers and eight hundred elementary and secondary school teachers.

The search for Jews, and for converted Jews, to be driven out of their jobs was continuous. On September 5, 1935 the SS newspaper published the names of eight half-Jews and converted Jews, all of the Evangelical-Lutheran faith, who had been "dismissed without notice" and deprived of any further opportunity "of acting as organists in Christian churches." From these dismissals, the newspaper commented, "It can be seen that the Reich Chamber of Music is taking steps to protect the church from pernicious influence."[3]

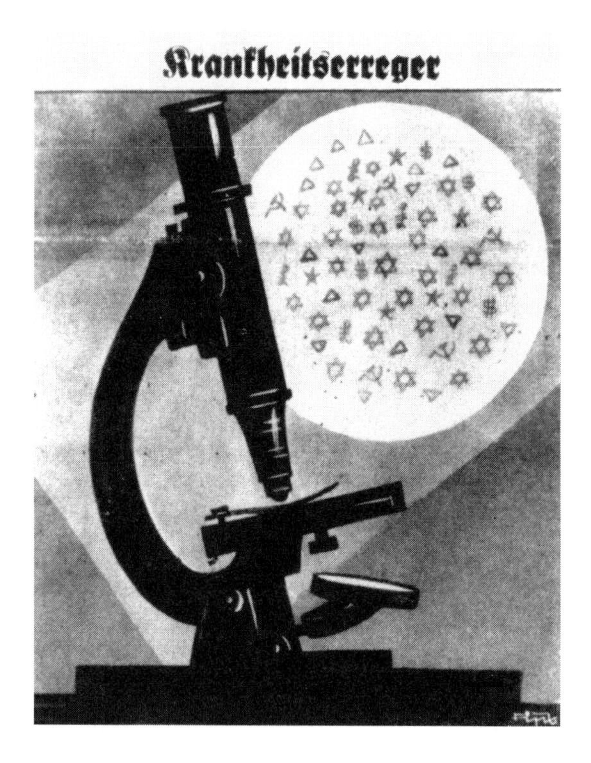

The illustration title reads "Infectious Germs." Under the microscope are symbols for Jews, communists, and homosexuals, along with symbols for the British pound and American dollar.

Little by little, antisemitism became a government policy. Jews and other "racial enemies" were singled out and then segregated and isolated. The next step would be annihilation. In time these same laws would be applied to "Gypsies" and Germans of African descent as well as Jews.

CONNECTIONS

What is the difference between an "inhabitant" and a "citizen"? How did that difference affect the way Germany defined its "universe of obligation"—the circle of individuals and groups toward whom obligations are owed, to whom rules apply, and whose injuries call for amends. What factors determined membership? Who was excluded? What were the consequences of being beyond the nation's "universe of obligation"?

The Nazis tried to find a racial definition of a Jew only to fail. As a result, they used religious practices to determine who was and was not a Jew. Earlier chapters detailed efforts in the United States to define an African American. Those efforts also failed. What questions might these failures have raised about the meaning of the term *race*? About its relevance to society?

As early as the 1910s, the Germans were aware of American anti-miscegenation laws. In the late 1930s, the Nazis noted that in many states in the United States, an individual with 1/32nd African ancestry was legally black. By contrast, individuals in Germany who were 1/8 Jewish were legally Aryans. What point were the Nazis trying to make? How valid was their argument?

Compare the "Law for the Protection of German Blood and German Honor" with Virginia's anti-miscegenation law (Chapter 6). What are the similarities? What differences do you notice?

---

1. *Documents on Nazism 1919-1945* ed. by Jeremy Noakes and Geoffrey Pridham. Schocken Books, 1983, 1984, pp. 463-467.
2. Ibid., p. 463.
3. *The Holocaust* by Martin Gilbert. Holt, 1985, p. 47.

Reading 7

In the summer of 1935, eugenicists, anthropologists, population scientists, and geneticists from all over the world traveled to Berlin, Germany, to take part in the International Congress for Population Science. Two Americans served as vice presidents of the conference: Harry Laughlin and Clarence Campbell. Although Laughlin was unable to attend, he wrote a paper for the conference and sent an exhibit consisting of 12 charts and publications that illustrated how the United States applied biological principles to its immigration policies.[1] Campbell not only attended but also publicly praised Nazi racial policies. He told delegates:

> The leader of the German nation, Adolf Hitler, ably supported by . . . the nation's anthropologists, eugenicists and social philosophers, has been able to construct a comprehensive racial policy of population development and improvement that promises to be epochal in racial history. It sets a pattern which other nations and other racial groups must follow if they do not wish to fall behind in their racial quality, in their racial accomplishments and in their prospects for survival. It is [true] that these ideas have met stout opposition in the . . . social philosophy which . . . bases its . . . whole social and political theory upon the patent fallacy of human equality. . . . But . . . human thought has not stood entirely still since the eighteenth century. [There is] a decided tendency . . . in enlightened minds no longer to place implicit faith in rhetorical principles which have no foundation in facts and to explore the realities of nature.
>
> Any patriotism worthy of the name carries with it a willingness on the part of individuals not only to cooperate in the common interest but to sacrifice individualistic aims and submit themselves to discipline in the ultimate interest of the group.
>
> A population group which is racially [uniform] and which has no racially alien elements which serve to confuse, obstruct and defeat its racial objectives will always tend to be unified in its racial objectives as well as have a high survival value and prospects.[2]

On his return to the United States, Campbell complained that the "anti-Nazi propaganda with which all countries have been flooded[has] gone far to obscure correct understanding and the great importance of the German race policy."[3] Like a number of other American eugenicists, Campbell dismissed reports of brutality toward the Jews as "Jewish propaganda"[4] at a time when the Nazis' campaign against the Jews was intensifying.

Even as Campbell defended the Nazis, thousands of Jews were trying desperately to leave the country. Some found sanctuary in various European countries. Others were unable to find a place to live. Everywhere they turned, they encountered barriers—not from Germany but from other nations. Adolf Hitler was eager to have Jews leave the country as long as they left their money behind. Few countries, however, were willing to accept thousands of penniless Jewish refugees. The barriers to entering the United States were especially high. In 1929, Congress amended the National Origins Act of 1924 to limit the number of immigrants who could enter the nation in a single year to 153,774. Each had to be in good health and of good character. Immigrants also had to prove that they were not likely to become "a public charge." Initially, American officials interpreted this to mean that families had enough money to tide them over until the adults found work—about $100, a considerable sum in the 1930s.

Every country had a number based on two percent of the total number of immigrants from that country living in the United States in 1890. As a result, 83,575 places were set aside for immigrants from Britain and Ireland. Germany had about 26,000 places; Poland, 6,000; Italy, 5,500; France, 3,000; and Romania, 300.

In 1930, in the midst of the Great Depression, President Herbert Hoover instructed the state department to issue visas only to applicants who were unlikely to ever become a public charge. Government officials interpreted the order to mean that a family had to have at least $10,000. As a result, immigration dropped sharply. Nearly 242,000 immigrants entered the United States in 1930. The number fell to 97,139 in 1931 and to 35,576 in 1932, the year before Hitler came to power. Of the 63,000 Jews who fled Germany between 1933 and 1934, only 6,514 were able to enter the United States. In contrast, France, a much smaller nation that was also in the midst of the Great Depression, accepted 30,000 Jewish refugees.

Even as the United States was raising the amount of money an immigrant needed to enter the nation, the Nazis were decreasing the amount Jews could take out of Germany. In January 1933, a Jew was allowed to take out as much as $10,000 in cash. The amount was reduced to $6,000, next to $4,000, then to $800, and finally in October of 1934 to $4 per immigrant.

In the early 1930s, as violence against Jews and other "racial enemies" increased in Germany, some Americans urged Congress to ease restrictions on immigration. They immediately encountered opposition led by John B. Trevor, the New York attorney who proposed the quota system in 1924. As head of the American Coalition of Patriotic, Civic and Fraternal Society, he asked Harry Laughlin to prepare a report on the effects of easing restrictions. In his report, Laughlin urged Congress to "offer no exceptional admission for Jewish refugees from Germany" and no admission to anyone without "a definite country to which he may be deported, if occasion demands," and anyone whose ancestors were not "all members of the white or Caucasian race."

Laughlin suggested that Congress "look upon the incoming immigrants, not essentially as in offering asylum nor in securing cheap labor," but primarily as "sons-in-law to marry their own daughters." In his view, "immigrants are essentially breeding stock." Lawmakers agreed.

CONNECTIONS

Campbell refers to the idea of human equality as a "patent fallacy." What ideas does he consider more important than equality? How do those ideas shape his definition of the word *patriotism*? The choices he made?

Not everyone at the conference applauded Campbell's speech. Two American scientists walked out of the conference. Another resigned from the Eugenics Research Association after returning to the United States. Why do you think some American eugenicists were now uncomfortable with Hitler's policies? How might Campbell's speech have contributed to their discomfort?

In the 1930s, how did American eugenicists like Laughlin define their universe of obligation? In their view who belonged? Who did not? How did the United States define its universe of obligation?

What are the consequences of an expanded universe of obligation? Of a very small universe of obligation? Who decides in a democracy where the lines will be drawn? Who decides in a dictatorship like Nazi Germany?

How did Laughlin view immigrants? Why did he seem to fear the newcomers?

Historian A. J. P. Taylor once wrote that Hitler took the Germans at their word. He made them "live up to their professions, or down to them--much to their regret." To what extent did Hitler also take American eugenicists at their word?

1. *The Nazi Connection* by Stefan Kühl.. Oxford University Press, 1994, p. 34.
2. Quoted in *The New York Times,* August 29, 1935.
3. Quoted in *The Nazi Connection* by Stefan Kühl. Oxford University Press, 1994, p. 35.
4. Quoted in *While Six Million Died* by Arthur D. Morse. Random House, 1967, p. 116.

Honorary Degrees and Propaganda

Reading 8

The Nazis rewarded American eugenicists whose work they admired. In 1936, Harry Laughlin of the Eugenics Record Office (ERO) received the following invitation from Carl Schneider, a professor of racial hygiene:

> The Faculty of Medicine of the University of Heidelberg intends to confer upon you the degree of Doctor of Medicine [by reason of honor] on the occasion of the 550-year Jubilee (27th to 30th of June 1936). I should be grateful to you if you could inform me whether you are ready to accept the honorary doctor's degree and, if so, whether you would be able to come to Heidelberg to attend the ceremony of honorary promotion and to personally receive your diploma.

The letter ended with a list of Laughlin's publications:

> A decade of progress in Eugenics. 1934
> Laughlin, A Report of the Special Committee on Immigr., 1934.
> " The Legal Status of Eugenical Sterilization,
> " Eugenical Sterilization in the United States, 1922.
> " Europe as an Immigrant-Exporting Cont., 1924.
> " Analysis of (the) America's Mod. Melting Pot, 1923.
> " Biological Aspects of Immigration, 1927.
> " Eugenical Aspects of Deportation, 1928.
> " Am. History in Terms of Human Migration, 1928
> " 21 Reprints.[1]

Laughlin promptly responded to the invitation:

> I stand ready to accept this very high honor. Its bestowal will give me particular gratification, coming as it will from a university deep rooted in the life history of the German people, and a university which has been both a reservoir and a fountain of learning for more than half a millennium. To me this honor will be doubly valued because it will come from a nation which for many centuries nurtured the human seed-stock which later founded my own country and thus gave basic character to our present lives and institutions.
>
> I regret more than I can say that the shortness of time before the jubilee date makes it impossible for me to arrange to leave my duties

at Cold Spring Harbor to visit Heidelberg to participate in the ceremony and to receive this highly honored diploma in person.[2]

Laughlin received his honorary degree at the German consulate in New York City. No one knows exactly why he decided not to travel to Germany to accept it. He may have been wary of an attack in the American press. *The New York Times* and other newspapers charged that Americans who traveled to Germany for such honors were being used as propaganda tools. Laughlin may have also feared that the effects of that kind of criticism on his relationship with the Carnegie Foundation, which was becoming more and more skeptical of his work.

Despite the criticism and worries about funding, Laughlin's belief in immigration restriction and the value of the Nazis' eugenics policies remained unshaken. Two years after he received his honorary degree, there was once again a move to allow Jewish refugees to enter the country. This time, the move was prompted by the violence that swept Germany and Austria on the night of November 10-11, 1938—*Kristallnacht* or the "Night of the Broken Glass" as it was later known. That night gangs of Nazis smashed, looted, and burned Jewish homes, businesses, and synagogues.

A month later, Laughlin reported on current projects to Wickliffe Preston Draper, a millionaire who had recently established the Pioneer Fund to fund eugenics outreach:

> You will be interested to know that the moving picture film "Eugenics in Germany" has proven very popular with senior high school students. Up to date the film has been loaned 28 times. Just now one copy is being used by the Society for Prevention of Blindness in New York, and the other is in the hands of George Smith. . . . where his advanced students in high school biology found it very interesting. Last spring Mr. Smith used the film with one set of students, and this year a second lot is profiting from it. . . . Most of the high schools now have projection apparatus so that films of this sort fit well into their program.[3]

Eugenics in Germany was a version of a Nazi propaganda film entitled "*Erbkrank*," or "The Genetically Diseased." After showing the entire movie at the Carnegie Institution in Washington, Laughlin secured funding from Draper's Pioneer Fund to distribute an edited version to the general public. Although the film depicts Jews as particularly susceptible to "hereditary degeneracy," Laughlin told readers of the *Eugenic News* that it contained "no racial propaganda of any sort."

The film was shown 28 times between 1937 and 1938, but plans to distribute it nationally fell through. Still the Nazis proclaimed the effort a great success. According to one German newspaper, the film made "an exceptionally strong impression" on American eugenicists.

CONNECTIONS

Why do you think the Nazis highlighted Laughlin's *Report of the Special Committee on Immigration,* 1934 in listing his major publications? Look carefully at the list of Laughlin's other publications. What appeal might they have for the Nazis?

Laughlin claimed that his degree would "be doubly valued because it will come from a nation which for many centuries nurtured the human seed—stock which later founded my own country and thus gave basic character to our present lives and institutions." What connection does he see between Germany and the United States? Who is part of that connection? Who is excluded?

Use the timeline on pages 251-254 to determine what Laughlin knew about Nazi Germany by 1936. To what extent did that knowledge influence his decision to accept the honorary degree? To what extent should that knowledge have influenced his decision? What did he know by the end of 1938? Why do you think his position remained unchanged despite the violence of *Kristallnacht*?

1. Schneider to Laughlin, May 16, 1936. Harry H. Laughlin Papers, Pickler Memorial Library, Truman State University, Kirksville, Missouri.
2. Laughlin to Schneider, May 28, 1936. Harry H. Laughlin Papers, Pickler Memorial Library, Truman State University, Kirksville, Missouri.
3. Courtesy of the Harry H. Laughlin Papers, Pickler Memorial Library, Truman State University, Kirksville, Missouri.

Reading 9

By the mid-1930s, Germany was a totalitarian state. The nation's courts, legislature, and other institutions were under Hitler's control. Individuals who spoke out against his regime were quickly silenced. Yet even in the United States, where the right to speak was protected by the Constitution, very few scientists were willing to take a stand. One of the few to do so was Franz Boas (Chapter 3).

Although he was 75 years old when the Nazis came to power in 1933, Boas, who once described a scientist as someone for whom the very essence of life is "the service of truth," argued that he and his colleagues were obligated to speak out as a community against race science, eugenics, and what he called "Nordic nonsense." Even though many American scientists privately agreed with his views, they were unwilling to take a public stand. When Boas asked Livingston Farrand, the president of Cornell University, to prepare a petition critical of German racism, Farrand refused. He argued that taking a public stand "as a rule does no good in a time of inflamed opinion and often delays understanding rather than aids it." Raymond Pearl, a former eugenicist, told Boas that scientists should not make statements on "political questions." In his view, petitions risked "harm to the scientific men who sign them and through these men to science itself." It was up to the German scientists, Pearl concluded, to take a stand, since Hitler was their leader.[1]

Harvard anthropologist E. A. Hooton was the only scientist willing to aid Boas, but not because he was opposed to eugenics. On the contrary, he had been a featured speaker at Charles Davenport's National Conference on Race Betterment. Still he rejected Nazi racism. At Boas's request, Hooton prepared a petition stating that there is no such thing as an "Aryan" or "Nordic race." "The so-called Nordic race is a hybrid . . . of several strains present in Europe during the post-glacial period," wrote Hooton. He also added that there is no scientific proof that some races are superior to others. Hooton sent the statement to seven anthropologists and asked that they join him in signing the document. Only one signed the petition.

Disillusioned but persistent, Boas continued to speak out against Nazi policies. Often, antisemitism hindered his efforts. At one point, he was nearly excluded from an important conference because the organizers feared that a Jew might be biased on questions of race. Yet those same organizers expressed no concerns about bias when they issued invitations to German scientists who actively supported Nazi policies.

In 1938, Boas and a few other like-minded scientists drafted yet another statement that challenged Nazi racial theories. By then, U.S. public opinion was beginning to turn against the Nazis. This time, about 50 leading scientists signed the document and others quickly followed suit. By October 1938, over one thousand scientists from across the United States had put their names on the statement. Even as Boas was gathering signatures, the Nazis were accelerating their campaign against the Jews. On November 10-11 came *Kristallnacht*, the "Night of Broken Glass." Although the violence directed against Jews that night did not alter Harry Laughlin's views, it had an enormous impact on other American scholars. By December 10, 1938, about 1,300 had signed Boas's statement. Later that month, the American Anthropological Association passed a resolution drafted by Hooton and introduced by Boas. The resolution read as follows:

> Whereas, the prime requisites of science are the honest and unbiased search for truth and the freedom to proclaim such truth when discovered and known, and
>
> Whereas, anthropology in many countries is being conscripted and its data distorted and misinterpreted to serve the cause of an unscientific racialism rather than the cause of truth:
>
> Be it resolved, that the American Anthropological Association repudiates such racialism and adheres to the following statement of facts:
>
> 1. Race involves the inheritance of similar physical variations by large groups of mankind, but its psychological and cultural connotations, if they exist, have not been ascertained by science.
> 2. The terms Aryan and Semitic have no racial significance whatsoever. They simply denote linguistic families.
> 3. Anthropology provides no scientific basis for discrimination against any people on the ground of racial inferiority, religious affiliation, or linguistic heritage.[2]

The following year, at the Seventh International Genetics Congress in Edinburgh, Scotland, a group of scientists prepared what became known as the Geneticists' Manifesto. It called for "the removal of race prejudices and of the unscientific doctrine that good or bad genes are the monopoly of particular peoples or persons with features of a given kind."[3]

Most scientists, however, were slow to challenge eugenicists. Scientist Jonathan Marks writes:

> Only well after the [eugenics] movement had been widely criticized by people outside of genetics and biology did the biologists begin to fall away from the movement. Possibly they were late to do

so because the eugenics movement was advancing the cause of genetics and biology in the America—which brought greater attention to the work biologists were doing and greater funding potential . . . . If biologists did in fact widely see the abuse to which genetic knowledge was being put, but refused to criticize it out of self-interest, they paid dearly for it. As historians of genetics have noted, the eugenics movement ultimately cast human genetics in such a disreputable light that its legitimate development was retarded for decades.[4]

CONNECTIONS

Compare and contrast the way Boas viewed the role of a scientist in society with the way Charles Davenport, Harry Laughlin, and other eugenicists viewed that role. What similarities do you notice? How important are the differences?

Why was Boas vulnerable to charges of bias? What other individuals or groups feel similarly vulnerable when they try to challenge prejudice? What justifications did Farrand and Pearl give for refusing to support Boas? How would you respond to the argument that speaking out when public opinion is "inflamed" does no good? To the idea that scientists should not become involved in "political questions"?

How courageous was Boas's stand? Why do you think so many other scientists and scholars were reluctant to join him in challenging Nazi ideas, even though in the United States they could do so in safety? If they had protested, would their words have had any effect in Nazi Germany? On American public opinion?

1. "Mobilizing Scientists Against Nazi Racism, 1933-1939" by Elazar Barkan in *Bones, Bodies, Behavior*, ed. by George Stocking, University of Wisconsin Press, 1988, p. 186.
2. Ibid., p. 202.
3. Quoted in *The Legacy of Malthus* by Allan Chase. Alfred A. Knopf, 1977, p. 614.
4. *Human Biodiversity: Genes, Race, and History* by Jonathan Marks. Aldine De Gruyter, 1995, p. 93.

Reading 10

A number of American scientists told Franz Boas that there was no need to speak out against eugenics because scientific discoveries were undercutting both eugenics and racism. In 1913, A. H. Sturtevant, a student of Thomas Hunt Morgan (Chapter 3), produced the first gene map. It showed that genes are located in a specific order on a chromosome. Gregor Mendel was mistaken in thinking that hereditary particles (genes) are always randomly arranged during reproduction. If Mendel had looked at traits associated with genes on the same chromosome, he might have discovered that his ratios of dominant to recessive traits do not work. Heredity is more complicated than he realized. Herman Muller, another student of Morgan's, found that X-rays can cause mutations in fruit flies. By showing that the physical environment can alter genes, it undercut the eugenic notion that genes are immune to outside influences.

Geneticists were also learning that repeated breeding within a so-called "pure" line does not lead to better specimens, as eugenicists predicted. Instead, it results in a general decline in health and hardiness. Because inbred strains lack genetic variation, they experience more hereditary defects. On the other hand, crossing strains leads to what scientists call "hybrid vigor." Such discoveries contradicted eugenic beliefs about "purity" and "superiority."

Logic also undermined eugenics. British geneticist Reginald Punnet questioned Henry Goddard's claim that sterilization would reduce feeblemindedness in the general population. Even if a recessive gene caused feeblemindedness (and it does not), Punnet noted that sterilization was unlikely to solve the problem. After all, a person can carry that gene without being feebleminded. How then would you decide whom to sterilize? Punnet concluded that "even under the unrealistic assumption that all the feebleminded could be prevented from breeding, it would take more than 8,000 years before their numbers were reduced to 1 in 100,000, given Goddard's estimate that about 3 in 1,000 Americans were genetically feebleminded.[1]

Partly in response to a growing skepticism about the value of eugenics as well to concerns about Hitler's "racial state," the Carnegie Foundation, which had long funded the Eugenics Record Office (ERO), asked a group of independent scholars to evaluate its work. In 1935, they described the ERO's research as "unsatisfactory for the study of genetics" and recommended that the group "cease from engaging in all forms of propaganda and the urging or sponsoring of programs for social reform or race betterment such as sterilization, birth control, inculcation of race or national consciousness, restriction of immigration, etc."[2] Even before the report was issued, the directors of the Carnegie Foundation persuaded

Charles Davenport to retire. In 1939, at their request, Harry Laughlin also resigned his post. Soon after, the ERO closed its doors.

At about the same time, many established scientists resigned from such eugenic organizations as the Galton Society and the American Eugenics Society. By the time the United States entered World War II in 1941, American eugenicists had broken all ties with the Nazis. Their organizations suspended most of their activities for the duration of the war. Yet the values and beliefs about difference that defined the movement did not disappear. They continued to appeal to many Americans long after the world confronted the consequences of Nazi eugenic racism.

Ironically, a project funded by the Carnegie Corporation in the mid-1930s reshaped discussions of race in the years after the war. It was a scientific study of race relations in the United States similar to the one Franz Boas asked Andrew Carnegie to fund in 1905. (See pages 87-88.) The new study was headed by Gunnar Myrdal, a Swedish sociologist who spent seven years gathering information about race in the United States. In 1944, just as the war was coming to a close, he published his findings in a book entitled *The American Dilemma.* Frederick Keppel, president of the Carnegie Corporation, wrote the foreword. It says in part:

> When the Trustees of the Carnegie Corporation asked for the preparation of this report in 1937, no one (except possibly Adolf Hitler) could have foreseen that it would be made public at a day when the place of the Negro in our American life would be the subject of greatly heightened interest in the United States. . . . The eyes of men of all races the world over are turned upon us to see how the power of the most powerful of the United Nations [is] dealing at home with a major problem of race relations.[3]

In the introduction to his book, Myrdal defined the "American dilemma":

> Though our study includes economic, social, and political race relations at the bottom our problem is the moral dilemma of the American—the conflict between his moral valuations on various levels of consciousness and generality. The "American Dilemma," . . . is the ever-raging conflict between, on the one hand, the valuations preserved on the general plane which we shall call the "American Creed," . . . and, on the other hand, the valuations on specific planes of individuals and groups living where personal and local interest, economic, social, and sexual jealousies [exist].[4]

CONNECTIONS

What scientific developments undermined the claims of eugenics? How did each, step by step, finding by finding, alter the way scientists viewed eugenics? Why do you think that few of these breakthroughs were publicized? What does your answer suggest about the way the media viewed the role of scientists in society?

According to Mrydal, what is the "American dilemma"? How would you define it? Does it still exist today? How is the nation trying to resolve it? How successful has the nation been? What does Myrdal's definition of the "American dilemma" suggest about the way he defines the nation's "universe of obligation"? Whom does he seem to include in the nation? To exclude?

---

1. "The Hidden Science of Eugenics" by Dianne Paul and Hamish Spencer. *Nature,* vol. 374, March 1995, p. 302.
2. "The Eugenic Record Office at Cold Spring Harbor, 1910-1940: An Essay in Institutional History" by Garland Allen. *Osiris*, 2nd series, 1986, vol. 2, p. 252.
3. "Foreword" by Frederick Keppel in *The American Dilemma: The Negro and Modern Democracy* by Gunnar Myrdal. Harper & Brothers, 1944.
4. *The American Dilemma: The Negro and Modern Democracy* by Gunnar Myrdal. Harper & Brothers, 1944, Introduction.

Reading 11

In the 1930s, Nazi policies forced not only scientists but also ordinary citizens to make choices. The Nazis did not turn Germany into a "racial state" all at once. The change took place step by step, decree by decree. Each new policy went a littler further than those enacted earlier. At each step, the German people had to make decisions. Yet even as they compromised and rationalized, few dared to ask, "Where is this path taking us?"

In the fall of 1933, a few months after the sterilization law took effect, Germany's Minister of Justice proposed a law that would allow "mercy killing" or euthanasia. Like the sterilization law, it was widely discussed in not only Germany but also the United States. *The New York Times* ran a front-page story about the proposal. It quoted a Nazi official who claimed the law would allow physicians "to end the tortures of incurable patients, upon request, in the interests of true humanity." The courts would decide who was incurable in much the way they determined who would be sterilized.[1] Although few people objected to the sterilization law, Catholic and Lutheran religious leaders were outraged at the idea of a "euthanasia" law. As a result, the proposal was quietly tabled.

Adolf Hitler did not give up on the idea, however. Throughout the 1930s, he

The poster shows how much the Prussian government provides annually for the following (left to right): a normal schoolchild, a slow learner, the educable mentally ill, and a blind or deaf-born schoolchild.

used propaganda to build support for the program by describing as "marginal human beings" epileptics, alcoholics, and individuals with birth defects, hearing losses, mental illnesses, and personality disorders, as well as those who were visually impaired or suffered from certain orthopedic problems. In 1936, the Nazis honored not only Harry Laughlin with an honorary degree but also Foster Kennedy, an American psychiatrist who proposed that "defective children" be "relieved of the agony of living."[2]

In the spring of 1939, as Germany prepared for war, Hitler set up a committee of physicians to prepare for the murder of disabled and "retarded" children. Known as the "Reich Committee for the Scientific Treatment of Severe Hereditary and Congenital Diseases," the group was told to keep its mission secret. Two weeks before the invasion of Poland in September of 1939, members asked physicians and midwives to fill out a questionnaire for every child born with a deformity or disability. The committee claimed that the data would be used "to clarify certain scientific questions." In fact, it was used to determine the fate of each child.

The committee never examined a single child, consulted with any youngster's physician, or spoke to relatives. Instead members used questionnaires to decide who would live and who would die. Once the decision was made, the child's parents were told only that the youngster was being placed in a special hospital to "improve" treatment. There death came quickly. The program was later expanded to include not only young children but also teenagers and adults. One "euthanasia expert" justified the murders by arguing, "The idea is unbearable to me that the best, the flower of our youth, must lose its life at the front, in order that feebleminded and asocial elements can have a secure existence in the asylum." Another suggested that a doctor has a duty is to rescue the "fit" for the future by weeding out the "unfit."

Although the program was kept secret, many Germans were aware of the killings. In some places, hundreds of individuals were murdered in a matter of days. Mobile gas vans carried out some of the killings. By June 1940, the vans were being replaced with "showers" that sprayed gas. Between 1939 and 1941 at least 70,000 persons were killed. A number of experts place the figure higher, claiming that at least 250,000 were murdered.

In November of 1940, Else von Löwis, a long-time supporter of Hitler and the Nazi party, wrote to a friend, the wife of the chief justice of the Nazi supreme court:

> Undoubtedly you know about the measure now used by us to dispose of incurable insane persons; still, perhaps you do not fully realize the manner and scope of this, nor the horror it creates in

> people's minds! Here, in Württemberg, the tragedy takes place in Grafeneck, on the Alb. . . . In the beginning one instinctively refused to believe the tale, or in any case considered the rumors to be extremely exaggerated. On the occasion of our last business meeting at the Gau School in Stuttgart, about the middle of October, I was still told by a "well-informed" person that this involved only idiots, strictly speaking, and that application of "euthanasia" applied only to cases which have been thoroughly tested. It is entirely impossible now to make anybody believe that version, and individual cases established with absolute certainty spring up like mushrooms. One might deduct perhaps 20 percent but if one tried to deduct 50 percent this would not help. . . .
>
> I am of the opinion that the people have the right to know about the law, just as they knew of the sterilization law. The most awful thing in the present case is "the public secret" which creates a feeling of uneasiness. . . . Those who are responsible for those measures, have no concept of the measure of confidence they have thereby destroyed. Everybody must at once ask: What then can still be believed? Where is this path taking us and where should the boundary line be established? . . . [3]

The judge passed the letter on to Heinrich Himmler who ordered the closing of the facility near von Löwis's home. He did not stop the program, however. It continued until May 1941, when the Reich Committee for the Scientific Treatment of Severe Hereditary and Congenital Diseases began sending questionnaires to homes for the elderly. A few months later, Clemens Graf von Galen, the Catholic bishop of Munster, asked his congregation, "Do you or I have the right to live only as long as we are productive?" If so, he argued, "Then someone has only to order a secret decree that the measures tried out on the mentally ill be extended to other 'nonproductive' people, that it can be used on those incurably ill with a lung disease, on those weakened by aging, on those disabled at work, on severely wounded soldiers. Then not a one of us is sure anymore of his life."[4] The sermon was secretly reproduced and distributed throughout Germany.

Three weeks later, Hitler signed an order officially ending the program. In fact, it continued secretly throughout the war and may have claimed 100,000 more lives. And the mobile vans and showers that released gas instead of water were later used at Auschwitz and other Nazi death camps as part of the Holocaust—Hitler's plan to murder all of Europe's Jews.

CONNECTIONS

Although most Nazi activities against the "other" were loudly proclaimed, the "euthanasia" program was kept secret. Why do you think the Nazis did this? Why do you think they waited until the nation was at war to implement the program?

A Nazi eugenics manual referred to physicians as "alert biological soldiers." What does the name mean? How does it redefine the role of a physician? Physicians are bound by the Hippocratic oath—a vow to help the sick and abstain from any act that may be harmful to the patient. What is the relationship of such a physician to his or her patients? How did the sterilization act alter that relationship? What changes did the "euthanasia program" bring to that relationship?

To what prejudices do the posters included in this reading appeal? How do they justify killings without ever mentioning them? How are they like the posters used at eugenics fairs? (Chapter 5) What differences seem most striking?

Why weren't Else von Löwis and her neighbors outraged at the discovery that the mentally ill were being murdered? How did she seem to define her "universe of obligation"? Who belongs and who does not? Where did she draw the line? Why was she uncomfortable with the idea of a "public secret"? Can something that everyone knows be a secret?

In 1944, rumors of the mass murder of Jews reached Berlin. There, too, people had to decide how to respond, where to draw the line. Ruth Andreas-Friedrich, a journalist who belonged to a resistance group, wrote in her diary:

> "They are forced to dig their own graves," people whisper. "Their clothing, shoes, shirts are taken from them. They are sent naked to their deaths." The horror is so incredible that the imagination refuses to accept its reality. Something fails to click. Some conclusion is not drawn. . . . We don't permit our power of imagination to connect the two, even remotely. . . . Is it cowardice that lets us think this way? Maybe! But then such cowardice belongs to the primeval instincts of man. If we could visualize death, life as it exists would be impossible. . . . Such indifference alone makes continued existence possible. Realizations such as these are bitter, shameful and bitter.[5]

Why does Andreas-Friedrich believe that "indifference alone makes continued existence possible"? Why does she describe that realization as "bitter" and "shameful"? Among the few Germans willing to act on the rumors were Hans

Scholl and his younger sister Sophie. Read their story in Chapter 8 of *Facing History and Ourselves: Holocaust and Human Behavior.* What does their story teach us about the consequences of indifference? How might they answer the question Andreas-Friedrich raises? How would you answer it?

In July 1942, the *American Journal of Psychiatry* published two articles, one in favor of killing severely retarded children and the other opposed to the idea. Foster Kennedy wrote the article in favor of the murder of "defective children." The editors expressed the opinion that in due time, euthanasia like sterilization would become an accepted practice in the United States. They even suggested a public education campaign to overcome resistance. It is very likely that Kennedy and the editors knew about the German program. A few years earlier, journalist William L. Shirer described much of it in his best-selling book, *Berlin Diary.* The portion of the book that dealt with the murder of the disabled was republished in the June 1941 issue of *Reader's Digest,* then the most widely read magazine in the nation. How do the editors and Foster Kennedy define their moral community or "universe of obligation"? Who belongs and who does not? Where do they seem to "draw the line"? The article in favor of killing retarded children did not result in a public outcry. For the most part, the essay was ignored. Why do you think few Americans expressed outrage at the idea?

After World War II finally ended, the Allies accused a small group of German racial hygienists of participating in government-sponsored massacres. In their defense, they pointed to the United States as proof that elimination of "inferior elements" was not unique to Germany. Karl Brandt, the head of the Nazi program for the killing of the mentally disabled, told the court that the Nazi program for the sterilization and elimination of "life not worthy of living" was based on ideas and experiences in the United States. How would you respond to that argument? Does it absolve Brandt and the others of wrongdoing?

---

1. Quoted in *By Trust Betrayed* by Hugh Gregory Gallagher. Holt, 1990, pp. 93-94.
2. Ibid., p. 94.
3. Ibid., pp. 154-155.
4. Quoted in *Nazism: A History in Documents and Eyewitness Accounts, 1919-1945,* vol. 2, ed. by J. Noakes and G. Pridham. Schocken Books, 1988, p. 1038.
5. *Berlin Underground* by Ruth Andreas-Friedrich. Trans. by Barrows Mussey. Holt, 1947, pp. 116-117.

Reading 12

Henry Wallace, the vice president of the United States from 1941 to 1945, was one of the few American politicians to challenge both Nazi racism and American eugenics. Like Harry Laughlin, Wallace came from Iowa. Like Laughlin, he studied agriculture and genetics in college. But Wallace's vision of science and his view of humanity were very different from Laughlin's. When Wallace served as secretary of agriculture in 1933, he brought not only scientific knowledge and skills to his work but also a compassion for the poor.

In 1939, Wallace spoke to a group of scientists in New York at a dinner to celebrate Abraham Lincoln's birthday. He dedicated his speech to anthropologist Franz Boas for his work in "marshaling the moral forces of science" in the defense of democratic freedom.

> The cause of liberty and the cause of true science must always be one and the same. For science cannot flourish except in an atmosphere of freedom, and freedom cannot survive unless there is a honest facing of facts. The immediate reason for this meeting is the profound shock you have had, and the deep feeling of protest that stirs in you, as you think of the treatment some of your fellow scientists are receiving in other countries. Men who have made great contributions to human knowledge and culture have been deprived of their positions and their homes, put into concentration camps, driven out of their native lands. Their life work has been reviled.
>
> In those same countries, other men, who call themselves scientists, have been willing to play the game of the dictators by twisting science into a mumbo-jumbo of dangerous nonsense. These men are furnishing pseudo-scientific support for the exaltation of one race and one nation as conquerors.
>
> These things run counter to your whole tradition as scientists. You are not only amazed and shocked and moved to protest against the fate of your fellow scientists abroad. You shudder with the realization that these things have happened in scientifically advanced countries in the modern world—and that they might happen here.
>
> Claims to racial superiority are not new in the world. Even in such a democratic country as ours, there are some who would claim that the American people are superior to all others. But never before in the world's history has such a conscious and systematic effort been made to inculcate the youth of a nation with ideas of racial

superiority as are being made in Germany today.

Just what are these ideas? Let me quote from a translation of the *Official Handbook for the Schooling of the Hitler Youth*, the organization which includes some seventy percent of all the boys and girls in Germany of eligible age.

The handbook discusses the various races found in Germany and other parts of Europe. Concerning what it calls the Nordic race, it says: "Now what distinguishes the Nordic race from all others? It is uncommonly gifted mentally. It is outstanding for truth and energy. Nordic men for the most part possess, even in regard to themselves, a great power of judgment. They incline to be taciturn and cautious. They feel instantly that too loud talking is undignified. They are persistent and stick to a purpose when once they have set themselves to it. Their energy is displayed not only in warfare but also in technology and scientific research. They are predisposed to leadership by nature."

But here is what the handbook says concerning what it calls the "Western race," found principally in England and France: "Compared to the Nordic race there are great differences in soul-qualities. The men of the Western race are . . . loquacious. In comparison with the Nordic . . . men they have much less patience. They act more by feeling than by reason. . . . They are excitable, even passionate. The Western race with all its mental excitability lacks creative power. This race has produced only a few outstanding men."

Thus the dictatorial regime in Germany, masquerading its propaganda in pseudo-scientific terms, is teaching the German boys and girls to believe that their race and their nation are superior to all others, and by implication that that nation and that race have a right to dominate all others.

When I was a small boy, George Carver, a Negro who is now a chemist at Tuskegee Institute, was a good friend of my father's at the Iowa State College. Carver at that time was specializing in botany, and he would take me along on some of his botanizing trips. It was he who first introduced me to the mysteries of botany and plant fertilization. Later on I was to have an intimate acquaintance with plants myself, because I spent a good many years breeding corn. Perhaps that was partly because this scientist, who belonged to another race, had deepened my appreciation of plants in a way I could never forget.

Carver was born in slavery, and to this day he does not definitely know his own age. In his work as a chemist in the South, he correctly sensed the coming interest in the industrial use of the products of the farm—a field of research which our government is now pushing. I mention Carver simply because he is one example of a truth of which we who meet here today are deeply convinced. Superior ability is not the exclusive possession of any one race or any one class. It may arise anywhere, provided men are given the right opportunities.

It is the fashion in certain quarters to sneer at those so-called "poor whites," who suffer from poor education and bad diet, and who live in tumble-down cabins without mattresses. And yet I wonder if any scientist would care to claim that 100,000 children taken at birth from these families would rank any lower in inborn ability than 100,000 children taken at birth from the wealthiest one percent of the parents of the United States. If both groups were given the same food, housing, education and cultural traditions, would they not turn out to have about equal mental and moral traits on the average? If 100,000 German babies were raised under the same conditions as 100,000 Hindu babies or 100,000 Jewish babies, would there be any particular difference? No such experiments have been made or are likely to be made and so no absolutely scientific answer can be given. But when I raise such a question, I mean to imply that every race, every nation, and people from every economic group of society are a great genetic mixture. There is far greater variability among the heredity of individuals within the groups than among the groups. There may be a certain amount of stability of type with regard to skin and eyes and hair, but with regard to mental and emotional characteristics there is very little evidence of genetic uniformity for any race or nation. There may be a great deal of uniformity with respect to traditions but not with respect to complex hereditary characters.

On the whole, it seems probable that nowhere in the world in the next couple of centuries will a genuinely scientific attempt, in the sense understood by the plant or animal breeder, be made to breed for superior types of human beings. The different races and nations will continue to be conglomerates with a vast variability of mental and emotional qualities and the other abilities which make for leadership and genius.

Under what conditions will the scientist deny the truth and pervert his science to serve the slogans of tyranny? Under what conditions are great numbers of men willing to surrender all hope of individual freedom and become ciphers of the State? How can these conditions be prevented from occurring in our country?

Seeking to answer all such questions honestly, we shall inevitably come upon certain truths that are not flattering to us. We shall find in our own country some of the conditions that have made possible what we see abroad. It is not enough simply to hope that these conditions will not reach such extremes here as they have in some other countries. We must see to it that they do not. When a political system fails to give large numbers of men the freedom it has promised, then they are willing to hand over their destiny to another political system. When the existing machinery of peace fails to give them any hope of national prosperity or national dignity, they are ready to try the hazard of war. When education fails to teach them the true nature of things, they will believe fantastic tales of devils and magic. When their normal life fails to give them anything but monotony and drabness, they are easily led to express themselves in unhealthy or cruel ways, as by mob violence. And when science fails to furnish effective leadership, men will exalt demagogues, and science will have to bow down to them or keep silent.

These are the conditions that made possible what we are now witnessing in certain large areas of the world. They are the seeds of danger to democracy. Given a healthy, vigorous, educated people, dignified by work, sharing the resources of a rich country, and sure that their political and economical system is amply meeting their needs—given this, I think we can laugh at any threat to American democracy. But democracy must continue to deliver the goods.

Let us dedicate ourselves anew to the belief that there are extraordinary possibilities in both man and nature which have not yet been realized, and which can be made manifest only if the individualistic yet co-operative genius of democratic institutions is preserved. Let us dedicate ourselves anew to making it possible for those who are gifted in art, science and religion to approach the unknown with true reverence, and not under the compulsion of producing immediate results for the glorification of one man, or group, one race or one nation.[1]

CONNECTIONS

Why do you think Wallace believes that the "cause of liberty and true science" must always be "one and the same"? What do they have in common?

Why does Wallace discuss the African American scientist George Washington Carver in his speech? How does his doing so undercut the notion of racial and biological determinism?

What are the "seeds of danger" in American democracy, according to Wallace? What role does he believe that scientists should play in the maintenance of democratic freedoms? Why does his see those who support dictators as "twisting science?" How does Wallace's vision of democracy differ from the one Charles Davenport describes in *Heredity in Relation to Eugenics*? How are they alike? How might Wallace respond to Davenport's statement?

Wallace quotes from a handbook for Hitler Youth. What is he suggesting about the power of education in general and textbooks in particular in "twisting science"? In promoting hatred?

---

1. Henry A. Wallace Papers, Special Collections Department, University of Iowa Libraries, Iowa City, Iowa.

9. Legacies and Possibilities

The humbling thing about science is that no matter how much you think you know, it is a certainty that the next generation will know more.

Terry B. Strom

Chapter 1 introduced the concepts central to this book by examining the idea of difference through various lenses, some fictional, others real. In subsequent chapters, those concepts were placed in a historical context. The history of racism and eugenics reveals the power of unexamined ideas to shape not only scientific research and public policy but also the daily lives of ordinary people. In the spring of 1945, as World War II came to an end, many individuals and groups confronted that power for the first time. Alan Moorhead, a British journalist, expressed the feelings of many people, when he wrote after inspecting a Nazi death camp, "With all one's soul, one felt: 'This is not war. Nor is it anything to do with here and now, with this one place at this one moment. This is timeless and all mankind is involved in it. This touches me and I am responsible. Why has it happened? How did we let it happen?'"

Those questions haunted political leaders, scientists, and ordinary citizens in the years that followed. In a documentary series entitled *The Ascent of Man*, scientist Jacob Bronowski reflected on the role of scientists in the Holocaust as he stood before the crematorium at Auschwitz—a death camp where members of his own family were murdered. He told viewers:

> It is said that science will dehumanize people and turn them into numbers. That is false, tragically false. Look for yourself. This is the concentration camp and crematorium at Auschwitz. This is where people were turned into numbers. Into this pond were flushed the ashes of some four million people. And that was not done by gas. It was done by arrogance. It was done by dogma. It was done by ignorance. When people believe that they have absolute knowledge, with no test in reality, this is how they behave. This is what men do when they aspire to the knowledge of gods.
>
> Science is a very human form of knowledge. We are always at the brink of the known, we always feel forward for what is to be hoped. Every judgment in science stands on the edge of error, and is personal. Science is a tribute to what we can know although we are fallible. In the end the words were said by Oliver Cromwell: "I beseech you, in the bowels of Christ, to think it possible you may be

mistaken." I owe it as a scientist to my friend Leo Szilard, I owe it as a human being to the many members of my family who died at Auschwitz, to stand here by the pond as a survivor and a witness. We have to cure ourselves of the itch for absolute knowledge and power. We have to close the distance between the push-button order and the human act. We have to touch people.[1]

How do we as individuals and as citizens cure the "itch for absolute knowledge and power"? How do we close the distance between the "push-button order and the human act"? This chapter explores such questions at a time when science is closer than ever to realizing Francis Galton's dream of "weeding out inferior traits and promoting superior qualities." Chapter 9 also helps us understand, as German historian Detlev J. K. Peukert once wrote, "The shadowy figures that look out at us from the tarnished mirror of history are—in the final analysis—ourselves."

The first two readings in this chapter return to the questions of Chapter 1: How do we as members of a society decide which differences matter and which do not? How do those decisions shape our ideas about what it means to be a human being in the 21st century? What is the role of a citizen in a modern, scientifically advanced society? The readings that follow apply those questions to current discussion on the relationship between science and society. Each of these readings is followed by suggestions for independent research or group projects. In tackling one or more of these investigations or designing one of your own, think carefully about what it means to be human in the world today. How do your ideas about humanity shape the way you define your role as a citizen in a democracy? How do they shape your values and beliefs? Why do you think scientists like physicist Leon M. Lederman frequently remind us that although science gives us a "powerful engine," in the end it is we who "steer the ship"? How can we best "steer that ship" at a time of truly revolutionary changes in science and medicine?

1. *The Ascent of Man* by Jacob Bronowski. Little, Brown, and Co., 1973, pp. 370, 374.

Reading 1

In 1940, in the midst of World War II, W. H. Auden reflected on the role of a citizen in a modern, scientifically advanced society in a poem he titled "The Unknown Citizen (To JS/07/M/378 This Marble Monument Is Erected by the State)."

He was found by the Bureau of Statistics to be
One against whom there was no official complaint,
And all the reports on his conduct agree
That, in the modern sense of an old-fashioned word, he was
saint,
For in everything he did, he served the Greater Community.
Except for the War until the day he retired
He worked in a factory and never got fired,
Yet he wasn't a scab or odd in his views,
For his Union reports that he paid his dues.
(Our report on his Union shows it was sound.)
And our Social Psychology workers found
That he was popular with his mates and liked a drink.
The Press are convinced that he bought a paper every day
And that his reactions to advertisement were normal in every
way.
Policies taken out in his name prove he was fully insured.
And his Health-card shows he was once in hospital but left it
cured.
Both Producers Research and High-Grade Living declare
He was fully sensible to the advantages of the Installment Plan
And had everything necessary to the Modern Man,
A phonograph, a radio, a car, and a frigidare.
Our researchers into Public Opinion are content
That he held the proper opinions for the time of year;
When there was peace, he was for peace; when there was
war, he went.
He was married and added five children to the population,
Which our Eugenicist says was the right number for a parent of
his generation.
And our teachers report that he never interfered with their
education.

Was he free? Was he happy? The question is absurd:
Had anything been wrong, we should certainly have heard.[1]

CONNECTIONS

Who holds power in Auden's "Greater Community"? What is the role of a citizen in that community? What is the relationship between science and society in that community? What is the role of a citizen in that "Greater Community"?

How would leaders in the American eugenics movement have answered the questions Auden asks at the end of his poem—"Was he free? Was he happy?" How might Jacob Bronowski (Introduction) answer them? How would you answer those questions? Why do you think Auden calls the questions "absurd"?

In the 1920s, a number of countries, including Britain, France, and the United States, built monuments to an "unknown soldier" who died in battle during World War I. For centuries, nations had built monuments to honor kings, generals, and other leaders. Now they went to great lengths to choose an anonymous warrior from the millions who died on the battlefields. How is a monument to an "unknown soldier" different from one that honors a particular individual? How is such a monument similar? Why has Auden chosen to honor an "unknown citizen"? What is the moral or lesson of Auden's "monument"?

In 1999, physicist Leon M. Lederman told a group of high school students, "Modern science, however abstract, is never safe. It can be used to raise mankind to new heights or literally to destroy the planet. . . . We give you a powerful engine. You steer the ship." Compare and contrast his definition of citizenship with the one Auden describes. What differences seem most striking? How would you describe the role of a citizen in the world today? List the attitudes and values that mark a "good citizen" in a democracy. Record your list in your journal. You may wish to revise or add to it as the chapter progresses.

---

1. "The Unknown Citizen" by W. H. Auden. Copyright © 1940 and renewed 1968 by W. H. Auden. From *Collected Shorter Poems.* Random House, 1968.

Reading 2

In 2000, scientists announced the completion of "the first survey of the entire human genome." That accomplishment brought science closer than ever to the kind of genetic engineering described in an episode of *Star Trek: The Next Generation* entitled "Masterpiece Society." (See summary on pages 31-32.) In that episode, the crew of the starship *Enterprise* visits Moab IV, a planet that has built a utopia much like the one Francis Galton, Charles Davenport, Harry Laughlin, and other eugenicists longed for. The crew's encounter raises important questions about what it means to be human. It also prompts reflection on the extent to which our genes decide our future.

Genes are the stretches of DNA that code for the structure of proteins. They are found in every cell of our body. To some, like the people of Moab IV, they are the "book of life"—they determine one's fate, one's destiny. Sociologist Barbara Katz Rothman is among those who disagree. She notes, "If genes are the 'book of life,' we have to realize that that book is constantly being written and rewritten by life." She explains:

> [Who] am I? . . . I'm a person in history, a person standing at a particular moment in time, living a life and trying to understand it.
>
> I'm a Jew who's just been to Germany again, to talk about prenatal testing and its possible eugenic consequences. The Germans are like children who've just touched a hot stove. Americans may talk cheerfully about how genomics is going to bring about medical revolutions, but Germans have a hard time using the language of genes and the language of politics in the same sentence without getting nervous.
>
> I'm a mother. . . . The son I gave birth to twenty-six years ago is gay. A "gay gene" would get me off the hook, loved ones have reassured me. It can't be my fault if it's "genetic." Fault? Is my son's sexuality an error that needs explaining, blame, forgiveness? Why a search for a gay gene? Are genes gay? Or are people? Or, actually, are people gay, or is gay just one of the ways of thinking about categorizing entire people based on parts of themselves? Ah, the complexities.
>
> I'm a white woman who's learned to function as a black mama: my youngest child, mine by adoption, is African American. What I thought I understood about the way race is constructed in America has been put to the test these past eleven years. There's a lot to be

said for "identity politics," for acknowledging that people learn from their actual experience in life. I'm not black—sometimes it surprises me when I look in a mirror and see I'm just as white as ever—but I'm a stakeholder in the black community in a way I was not before.

I'm still the child whose daddy died of cancer, and the woman whose stepfather did the same. I have all the cancer fears of anyone in this society and then some. I know all the warning signs and see warnings where there are no signs at all.

I'm a sociologist, trained and educated to avoid reductionism in all things. Social systems aren't just the people who make them up; they have rules and characteristics of their own, things you can't understand by looking only at individuals. Trees stand still, I remind my introductory sociology students: trees are very geographically stable life forms. A tree will be just where you left it. Forests move. Looked at over time, forests move across the face of a continent. Each tree lives or dies just where it is, and the whole moves. A whole is not just the sum of its parts. A person is not just the sum of his or her genes.

Like you, like everybody, I'm very complicated, filled with contradictions, stories, memory, and history. I'm more than my DNA, more than a collection of proteins. And I'm bothered, worried, saddened, sometimes frightened by a metaphor for personhood that sees us as just "information." My concerns, and yours, about the new genetics are not just some ethical obstacles to be overcome so that they can go ahead and cure cancer and all that. What we're concerned about here is not just how much of who and what we are is predetermined in a set of codes for proteins. What is at issue is what it means to be a person, and how we can live our lives as individuals, as families, and as communities of people.[1]

CONNECTIONS

Watch "Masterpiece Society" again or reread the summary on pages 31-32. How do the people of Moab IV decide which differences matter and which do not? How have those decisions shaped their understanding of what it means to be human? How might other eugenicists have regarded their choice? How does Barbara Rothman view them? How has your study of the American eugenics movement affected the way you would regard them? What has your study taught you about the consequences of such choices?

Barbara Rothman describes herself as "a person in history." How does she describe the complications of life at this "particular moment in time"? What does she suggest that we can learn from history about the forces and the choices that brought us to this moment? What does she suggest about the power of the ideas that energize social and religious movements?

Rothman asks, "Are genes gay? Or are people? Or, actually, are people gay, or is gay just one of the ways of thinking about categorizing entire people based on parts of themselves?" How would you answer her questions? What do your answers suggest about the way societies determine which differences matter?

Rothman teaches her students that trees stand still but forests move. What idea is she trying to convey? How does it apply to your study of racism and eugenics? If the whole is not just the sum of its parts, what is the relationship between the whole and its parts? What is the relationship, for example, between the individual and society? Between a person and his or her genes?

Rothman describes Germans as nervous about the political implications of genomics. Benno Müller-Hill, a molecular biologist at the Institute of Genetics at the University of Cologne, is among those Germans. He wrote in 1993:

> The German human geneticists . . . abandoned their patients to criminal politicians. . . . Can it happen again? Certainly not the way it happened then. But I think there is another, more modern way to abandon patients. If genetic differences lead to drastic differences in insurance rates and employment, the human geneticists who have discovered genotypes and all other geneticists will be accused of not having stopped this process to create a genetic "under-race." Certainly the circumstances will differ drastically from those in Germany. No Führer will be responsible. It will be the market place with all its participants that will possibly create such an outrage.[2]

What similarities does Müller-Hill see between the past and the present? What differences does he identify? How important are those differences? What do his concerns suggest about the relationship between science and politics? About the relationship between science and economics?

---

1. Barbara Katz Rothman, *The Book of Life: A Personal and Ethical Guide to Race, Normality, and the Implications of the Human Genome Project.* Beacon Press, 1998, 2001, pp. xiv-xv.
2. "Human Genetics in Nazi Germany" by Benno Müller-Hill in *Medicine, Ethics, and the Third Reich* edited by John J. Michalczyk. Sheed & Ward, 1994, p. 34.

Reading 3

Eugenicists divided the world into "races" and insisted that some "races" (their own in particular) were superior to others. In the early 1900s, the American Museum of Natural History in New York City organized its exhibitions around eugenic principles. In 2001, the museum opened a new exhibit entitled "The Genomic Revolution." In a prominent place, the organizers featured the following statement:

> **The Only Race Is Human Race**
> **No Biological Basis for Race**
>
> New data from the mapping of the human genome reveal that all humans are incredibly similar—in fact, we are 99.9% genetically identical. **We are all members of one species,** Homo sapiens. Scientists have confirmed, as they long suspected, **that there is no genetic or biological basis for race.**
>
> Genetic variation between people within the same "racial" group can be greater than the variation between people of two different groups. Many people of African descent are no more similar to other Africans than they are to Caucasians. Genetic distinctions between Asians and Caucasians are less pronounced than those between groups from, for example, parts of East and West Africa.
>
> No matter how scientists today scrutinize a person's genes, they can't determine with certainty whether an individual is from one "racial" group or another. **Differences of culture and society** distinguish one group from another, but these distinctions are not rooted in biology.
>
> "Mapping the DNA sequence variation in the human genome holds the potential for promoting the fundamental unity of all humankind." —Dr. Harold P. Freeman[1]

A number of museums and scholarly associations have issued similar statements. Yet, writes physiologist Jared Diamond, most people regard the existence of race as obvious, a matter of common sense. He explains:

> Our eyes tell us that the Earth is flat, that the sun revolves around the Earth, and that we humans are not animals. But we now ignore that evidence of our senses. We have learned that our planet is in fact round and revolves around the sun, and that humans are slightly modified chimpanzees. The reality of human races is another

commonsense "truth" destined to follow the flat Earth into oblivion.

The commonsense view of races goes somewhat as follows. All native Swedes differ from all native Nigerians in appearance: there is no Swede whom you would mistake for a Nigerian, and vice versa. Swedes have lighter skin than Nigerians do. They also generally have blond or light brown hair, while Nigerians have very dark hair. Nigerians usually have more tightly coiled hair than Swedes do, dark eyes as opposed to eyes that are blue or gray, and fuller lips and broader noses.

In addition, other Europeans look much more like Swedes than like Nigerians, while other peoples of sub-Saharan Africa—except perhaps the Khoisan peoples of southern Africa—look much more like Nigerians than like Swedes. . . .

What could be more objective?

As it turns out, this seemingly unassailable reasoning is not objective. There are many different, equally valid procedures for defining races, and those different procedures yield very different classifications. . . .

To understand how . . . uncertainties in classification arise, let's steer clear of humans for a moment and instead focus on [animals], about which we can easily remain dispassionate. Biologists begin by classifying living creatures into species. A species is a group of populations whose individual members would, if given the opportunity, interbreed with individuals of other populations of that group. But they would not interbreed with individuals of other species that are similarly defined. Thus all human populations, no matter how different they look, belong to the same species because they do interbreed and have interbred whenever they have encountered each other. Gorillas and humans, however, belong to two different species because—to the best of our knowledge—they have never interbred, despite their coexisting in close proximity for millions of years. . . .

How does that variability of traits by which we classify races come about in the first place?

Many geographically variable human traits evolved by natural selection to adapt humans to particular climates or environments. . . . Good examples are the mutations that people in tropical parts of the Old World evolved to help them survive malaria, the leading infectious disease of the Old-World tropics. One such mutation is the sickle-cell gene, so-called because the red blood cells of people with that mutation tend to assume a sickle shape. People bearing the gene are more resistant to malaria than people without it. Not surprisingly, the gene is absent from northern Europe, where malaria is nonexistent, but it's common in tropical Africa, where malaria is widespread. Up to 40 percent of Africans in such areas carry the sickle-cell gene. It's also common in the malaria-ridden Arabian Peninsula and southern India, and rare or absent in the southernmost parts of South Africa, among the Xhosas, who live mostly beyond the tropical geographic range of malaria.

The geographic range of human malaria is much wider than the range of the sickle-cell gene. As it happens, other antimalarial genes take over the protective function of the sickle-cell gene in malarial Southeast Asia and New Guinea and in Italy, Greece, and other warm parts of the Mediterranean basin. Thus human races, if defined by antimalarial genes, would be very different from human races as traditionally defined by traits such as skin color. As classified by antimalarial genes (or their absence), Swedes are grouped with Xhosas but not with Italians or Greeks. Most other peoples usually viewed as African blacks are grouped with Arabia's "whites" and are kept separate from the "black" Xhosas.

Antimalarial genes exemplify the many features of our body chemistry that vary geographically under the influence of natural selection. Another such feature is the enzyme lactase, which enables us to digest the milk sugar lactose. . . . Until about 6,000 years ago most humans, like all other mammal species, lost the lactase enzyme on reaching the age of weaning. The obvious reason is that it was unnecessary—no human or other mammal drank milk as an adult. Beginning around 4000 B.C., however, fresh milk obtained from domestic mammals became a major food for adults of a few human populations. Natural selection caused individuals in these populations to retain lactase into adulthood. Among such peoples are northern and central Europeans, Arabians, north Indians, and several milk-drinking black African peoples, such as the Fulani of West Africa. Adult lactase is much less common in southern European populations and in most other African black populations, as well as in all populations of East Asians, aboriginal Australians, and American Indians. . . .

Other visible traits that vary geographically among humans evolved by means of sexual selection. We all know that we find some individuals of the opposite sex more attractive than other individuals. We also know that in sizing up sex appeal, we pay more attention to certain parts of a prospective sex partner's body than to other parts. Men tend to be inordinately interested in women's breasts and much less concerned with women's toenails. Women, in turn, tend to be turned on by the shape of a man's buttocks or the details of a man's beard and body hair, if any, but not by the size of his feet. . . .

There is a third possible explanation for the function of geographically variable human traits, besides survival or sexual selection—namely, no function at all. A good example is provided by fingerprints, whose complex pattern of arches, loops, and whorls is determined genetically. Fingerprints also vary geographically: for example, Europeans' fingerprints tend to have many loops, while aboriginal Australians' fingerprints tend to have many whorls.

If we classify human populations by their fingerprints, most Europeans and black Africans would sort out together in one race, Jews and some Indonesians in another, and aboriginal Australians in still another. But those geographic variations in fingerprint patterns possess no known function whatsoever. They play no role in survival. . . . They also play no role in sexual selection. . . .

You've probably been wondering when I was going to get back to skin color, eye color, and hair color and form. After all, those are the traits by which all of us members of the lay public, as well as traditional anthropologists, classify races. Does geographic variation in those traits function in survival, in sexual selection, or in nothing?

The usual view is that skin color varies geographically to enhance survival. Supposedly, people in sunny, tropical climates around the world have genetically dark skin, which is supposedly analogous to the temporary skin darkening of European whites in the summer. The supposed function of dark skin in sunny climates is for protection against skin cancer. . . .

Alas, the evidence for natural selection of skin color dissolves under scrutiny. Among tropical peoples, anthropologists love to stress the dark skins of African blacks, people of the southern Indian peninsula, and New Guineans and love to forget the pale skins of Amazonian Indians and Southeast Asians living at the same latitudes. To wriggle out of those paradoxes, anthropologists then plead the excuse that Amazonian Indians and Southeast Asians may not have been living in their present locations long enough to evolve dark skins. However, the ancestors of fair-skinned Swedes arrived even more recently in Scandinavia, and aboriginal Tasmanians were black-skinned despite their ancestors' having lived for at least the last 10,000 years at the latitude of Vladivostok.

Besides, when one takes into account cloud cover, peoples of equatorial West Africa and the New Guinea mountains actually receive no more ultraviolet radiation or hours of sunshine each year

> than do the Swiss. Compared with infectious diseases and other selective agents, skin cancer has been utterly trivial as a cause of death in human history, even for modern white settlers in the tropics. . . .
>
> It wouldn't surprise me if dark skins do eventually prove to offer some advantage in tropical climates, but I expect the advantage to turn out to be a slight one that is easily overridden. But there's an overwhelming importance to skin, eye, and hair color that is obvious to all of us—sexual selection. . . .
>
> We all know how those highly visible "beauty traits" guide our choice of sex partners. Even the briefest personal ad in a newspaper mentions the advertiser's skin color, and the color of skin that he or she seeks in a partner. Skin color, of course, is also of overwhelming importance in our social prejudices. If you're a black African American trying to raise your children in white U.S. society, rickets and overheating are the least of the problems that might be solved by your skin color. [2]

In reflecting on his argument, Diamond notes, "Depending on whether we classified ourselves by antimalarial genes, lactase, fingerprints, or skin color, we could place Swedes in the same race as either Xhosas, Fulani, the Ainu of Japan, or Italians." He goes on to explain that the classifications we traditionally use are related to sexual selection. He finds that choice not surprising:

> These traits are not only visible at a distance but also highly variable; that's why they became the ones used throughout recorded history to make quick judgments about people. Racial classification didn't come from science but from the body's signals for differentiating attractive from unattractive sex partners, and for differentiating friend from foe.
>
> Such snap judgments didn't threaten our existence back when people were armed only with spears and surrounded by others who looked mostly like themselves. In the modern world, though, we are armed with guns and plutonium, and we live our lives surrounded by people who are much more varied in appearance. The last thing we need now is to continue codifying all those different appearances into an arbitrary system of racial classification.[3]

When her family spent six months in the Netherlands, Barbara Katz Rothman discovered how arbitrary racial classifications are. Fearful that her then five-year-old daughter Victoria would be the only "black kid in her class," Rothman was told her concerns were unfounded. Yet, Rothman writes:

She was the only black kid in her class. She was the only black kid I saw anywhere in that school. If I hadn't been reassured by people I genuinely like and trust, I'd have just been angry. As it was, I was puzzled. I walked over to a wall of photographs of the school going back for years and years, group after group of class photos. No black kids. I didn't say anything, just kept watching, thinking about it. A few days later, light dawned for me: there were dark-skinned kids from India and Pakistan in all the classes. Black kids. European-style black kids.

For an American, with an American sensibility of race, Indian and African kids are not both "black." For a Dutch person, with a different race system in his head, these were all black kids.

So what does that story prove, anyway? That the Dutch draw a different line? Maybe between the Dutch and everyone else? Not being Dutch, are all the blacks, well, black? The Indian kids in her class could see what my kid and I could see, the distinctiveness of African features over and above the similarity of skin color.

So does the story tell us that race is a socially constructed category, constructed differently in different places? Or does it tell us that the Dutch draw their lines so tightly around themselves that they don't bother to make finer discriminations—not that they don't see or experience the distinction as existing, but that they don't see why it should matter.

And is that what white Americans do when they see a black kid whose family has been in the United States since slavery days, a black kid whose family arrived two generations ago from Haiti, and a black kid who just immigrated here from Nigeria, and calls them all "African American," seeing no meaningful differences?[4]

Rothman explains:

People certainly do see race. We see race as this physical reality, this recognizable pattern of differences between people. It is foolish to try to persuade people that the differences don't exist. They do. It is pointless to try to convince people that the differences don't matter. They do.

What confuses us is that the differences exist physically, but matter socially. There are physical differences, and even physical consequences. But there is not a physical cause-and-effect relationship between them. Take something relatively simple: There is a much higher infant mortality rate among blacks than among whites in America. The differences between black and white women are there, real and measurable. But those differences, the physical, biological characteristics marked as race—level of melanin in the skin, shape of the nose, or whatever—are not the cause of the different infant mortality rates. The darkness of the mother is a physical, biological phenomenon, as is the death of the baby. But the relationship between the two is a social reality; it is the social consequence of race that causes the physical reality of death.[5]

CONNECTIONS

The organizers of the exhibition at the American Museum of Natural History placed a number of sentences and phrases in their statement on race in large and/or very dark type. Why do you think they chose to highlight those ideas? If you were to highlight Jared Diamond's essay in a similar way, which sentences or phrases would you emphasize? Compare your choices with those of your classmates. How do you account for similarities and differences?

To what extent is seeing believing? How does Jared Diamond challenge that idea? To what extent does Rothman's story challenge it? How does our culture shape what we see and what we fail to notice? How does culture affect the importance we place on the differences we see in the world?

Law professor Martha Minow writes, "When we identify one thing as unlike the others, we are dividing the world; we use our language to exclude, to distinguish—to discriminate." How do her comments apply to popular ideas about race? To the use of "racial categories" in everyday life? How do those categories affect the way we see ourselves? The way others view us?

One goal of education is to expose individuals to other ideas so that they can weigh alternatives and make wise decisions. What role can education play in ending the "social reality" of race? In small groups, brainstorm ideas for altering or abolishing harmful stereotypes. Report to the class on the idea or combination of ideas your group considers most effective in ending discrimination.

Find out how institutions in your community address the "social reality of race." What successes have you uncovered? What problems remain?

---

1. *www.amnh.org/exhibtions/genomics/1 identity/ninety nine.html*
2. "Race without Color" by Jared Diamond. *Discover,* November, 1994, p. 82-89.
3. Ibid., p. 89.
4. Barbara Katz Rothman, *The Book of Life: A Personal and Ethical Guide to Race, Normality, and the Implications of the Human Genome Project.* Beacon Press, 1998, 2001, pp. 51-52.
5. Ibid., p. 63.

Reading 4

In the early 1900s, race was the lens through which many Americans viewed the world. It was a lens that shaped people's ideas about who belongs and who does not. During those years, a few people resisted the laws and customs that supported the notions that regarded African Americans as "inferior." Little by little, they chipped away at segregation. Then, on May 17, 1954, in *Brown v. the Board of Education*, the United States Supreme Court ruled unanimously that separate public schools for black and white children were not and could never be equal. In communities across the nation, educators made plans to integrate their schools.

In the fall of 1957, officials in Little Rock, Arkansas, decided to integrate the schools gradually beginning with Central High School. That September, the arrival of nine African American students resulted in a year of protests and violence followed by the closing of every high school in the city for one year.

Forty years later, the once all-white student body at Central High was 58 percent black and 39 percent white. Much as it was forty years earlier, the school was still known for producing many of the state's brightest students. Those students were both black and white and many of them were later admitted to the nation's most prestigious universities. Yet at Central High School, the honors classes were predominately white and the regular classes primarily African American. No one seems sure why this was so. Some think it was a result of racism. Others attributed it to the poor academic preparation of incoming black students. There was a similar gap between the scores of black and white students on the SATs and other tests that measure intelligence. That gap, which exists in many communities, has troubled many scholars including Claude D. Steele, the Lucie Stern Professor in the Social Sciences at Stanford University. He writes:

> Over the past four decades African-American college students have been more in the spotlight than any other American students. . . . These students have borne much of the burden for our national experiment in racial integration. And to a significant degree the success of the experiment will be determined by their success.
>
> Nonetheless, throughout the 1990s the national college-dropout rate for African Americans has been 20 to 25 percent higher than that for whites. Among those who finish college, the grade-point average of black students is two thirds of a grade below that of whites.
>
> A recent study by William Bowen and Derek Bok, reported in their book *The Shape of the River*, brings some happy news: despite

this underachievement in college, black students who attend the most selective schools in the country go on to do just as well in postgraduate programs and professional attainment as other students from those schools. . . . Still, the underperformance of black undergraduates is an unsettling problem, one that may alter or hamper career development, especially among blacks not attending the most selective schools.

Attempts to explain the problem can sound like a debate about whether America is a good society, at least by the standard of racial fairness, and maybe even about whether racial integration is possible. It is an uncomfortably finger-pointing debate. Does the problem stem from something about black students themselves, such as poor motivation, a distracting peer culture, lack of family values, or . . . genes? Or does it stem from the conditions of blacks' lives: social and economic deprivation, a society that views blacks through the lens of diminishing stereotypes and low expectations, too much coddling, or too much neglect?

In recent years this debate has acquired a finer focus: the fate of middle-class black students. Americans have come to view the disadvantages associated with being black as disadvantages primarily of social and economic resources and opportunity. This assumption is often taken to imply that if you are black and come from a socio-economically middle-class home, you no longer suffer a significant disadvantage of race. . . .

But virtually all aspects of underperformance—lower standardized-test scores, lower college grades, lower graduation rates—persist among students from the African-American middle class. This situation forces on us an uncomfortable recognition: that beyond class, something racial is depressing the academic performance of these students.[1]

As Steele and his colleagues investigated the gap, they wondered if the underperformance of African American students was affected by what they called "stereotype threat"—"the threat of being viewed through the lens of a negative stereotype, or the fear of doing something that would inadvertently confirm that stereotype." Steele, an African American, believes a black student is more likely than other Americans to wonder whether his or her "race" will set boundaries to experiences and relationships. Steele explains:

With time he may weary of the extra vigilance these situations require. . . . To reduce this stress he may learn to care less about the situations and activities that bring it about—to realign his self-regard so that it no longer depends on how he does in the situation. We have called this psychic adjustment "disidentification." Pain is lessened by ceasing to identify with the part of life in which the pain occurs. This withdrawal of psychic investment may be supported by other members of the stereotype-threatened group—even to the point of its becoming a group norm. But not caring can mean not being motivated. And this can have real costs. When stereotype threat affects school life, disidentification is a high price to pay for psychic comfort. Still, it is a price that groups contending with powerful negative stereotypes about their abilities—women in advanced math, African-Americans in all academic areas—may too often pay.

Steele and his colleagues designed a series of experiments to test their ideas. As part of the first set of experiments, they statistically matched in ability level two groups of Stanford students, one black and one white. The students, one at a time, were asked to take a thirty-minute test made up of items from the advanced Graduate Record Examination in literature. Because the students were mainly sophomores, they all found the test difficult. The test was presented to students in two ways: as a test of ability or a laboratory task to find out how certain problems are solved. The results seemed to confirm Steele's hypothesis: "When the difficult verbal test was presented as a test of ability, black students performed dramatically less well than white students, even though we had statistically matched the two groups in ability level. Something other than ability was involved; we believed it was stereotype threat."

Steele writes of his experiment:

> In matters of race we often assume that when a situation is objectively the same for different groups, it is experienced in the same way by each group. This assumption might seem especially reasonable in the case of "standardized" cognitive tests. But for black students, difficulty with the test makes the negative stereotype relevant as an interpretation of their performance, and of them. They know that they are especially likely to be seen as having limited ability. Groups not stereotyped in this way don't experience this extra intimidation. And it is a serious intimidation, implying as it does that they may not belong in walks of life where the tested abilities are important—walks of life in which they are heavily invested. Like many pressures, it may not be experienced in a fully conscious way, but it may impair their best thinking.

Steele wondered if "the effects of stereotype threat come entirely from the fear of being stereotyped" or "from something internal to black students—self-doubt, for example." This time, he and his colleagues tested "white male students who were strong in math." Half were told that a difficult math test they were about to take was one on which "Asians generally did better than whites." The other half was simply told that the test was difficult. Steele reasoned that "if stereotype threat alone—in the absence of any internalized self-doubt—was capable of disrupting test performance, then white males taking the test after this comment should perform less well than white males taking the test without hearing the comment." That is just what happened. The results of related tests to measure the effects of gender and class stereotypes seemed to confirm Steele's findings.

Steele and his colleagues also discovered that "the most achievement-oriented students, who were also the most skilled, motivated, and confident, were the most impaired by stereotype threat." Steele explains why:

> A person has to care about a domain in order to be disturbed by the prospect of being stereotyped in it. . . . When we tested participants who identified less with these domains, what had been under our noses hit us in the face. None of them showed any effect of stereotype threat whatsoever.
>
> These weakly identified students did not perform well on the test: once they discovered its difficulty, they stopped trying very hard and got a low score. But their performance did not differ depending on whether they felt they were at risk of being judged stereotypically.

What can be done to overcome the "stereotype threat"? Steele believes that "the success of black students may depend less on expectations and motivation—things that are thought to drive academic performance—than on trust that stereotypes about their group will not have a limiting effect in their school world." To test this idea, Steele and his colleagues decided to find out whether boosting a student's self confidence before a test affected his or her score. It did not. He explains:

> What did raise the level of black students' performance to that of equally qualified whites was reducing stereotype threat—in this case by explicitly presenting the test as racially fair. When this was done, blacks performed at the same high level as whites even if their self-confidence had been weakened by a prior failure.
>
> These results suggest something that I think has not been made clear elsewhere: when strong black students sit down to take a difficult standardized test, the extra apprehension they feel in comparison with whites is less about their own ability than it is about having to perform on a test and in a situation that may be primed to treat them

stereotypically. We discovered the extent of this apprehension when we tried to develop procedures that would make our black participants see the test as "race-fair." It wasn't easy. African Americans have endured so much bad press about test scores for so long that, in our experience, they are instinctively wary about the tests' fairness. We were able to convince them that our test was race-fair only when we implied that the research generating the test had been done by blacks. When they felt trust, they performed well regardless of whether we had weakened their self-confidence beforehand. And when they didn't feel trust, no amount of bolstering of self-confidence helped.

In reflecting on how a school or a teacher can foster trust across the "racial divide," Steele and his colleagues set up yet another experiment. They invited black and white Stanford students to write essays about favorite teachers for possible publication in a journal. Before each student left the first writing session, a researcher took a Polaroid snapshot of the student and placed it on top of his or her essay for use "if the essay was published." The purpose was to let essay writers know that the person evaluating their writing was aware of their race. Steele describes what happened when the writers received feedback on their work:

> We found that neither straight feedback nor feedback preceded by the "niceness" of a cushioning statement ("There were many good things about your essay") was trusted by black students. They saw these criticisms as probably biased, and they were less motivated than white students to improve their essays. White students took the criticism at face value—even as an indication of interest in them. Black students, however, faced a different meaning: the "ambiguating" possibility that the criticism was motivated by negative stereotypes about their group as much as by the work itself. Herein lies the power of race to make one's world insecure—quite apart from whatever actual discrimination one may experience.
>
> But this experiment also revealed a way to be critical across the racial divide: tell the students that you are using high standards (this signals that the criticism reflects standards rather than race), and that your reading of their essays leads you to believe that they can meet those standards (this signals that you do not view them stereotypically). This shouldn't be faked. High standards, at least in a relative sense, should be an inherent part of teaching, and critical feedback should be given in the belief that the recipient can reach those standards. These things go without saying for many students. But they have to be

made explicit for students under stereotype threat. The good news of this study is that when they are made explicit, the students trust and respond to criticism. Black students who got this kind of feedback saw it as unbiased and were motivated to take their essays home and work on them even though this was not a class for credit. They were more motivated than any other group of students in the study—as if this combination of high standards and assurance was like water on parched land, a much needed but seldom received balm. . . .

My colleagues and I believed that our laboratory experiments had brought to light an overlooked cause of poor college performance among non-Asian minorities: the threat to social trust brought about by the stereotypes of the larger society. But to know the real-life importance of this threat would require testing . . . in the buzz of everyday life.

To this end [we] undertook a program aimed at incoming first-year students at the University of Michigan. Like virtually all other institutions of higher learning, Michigan had evidence of black students' underachievement. Our mission was clear: to see if we could improve their achievement by focusing on their transition into college life. We also wanted to see how little we could get away with—that is, to develop a program that would succeed broadly without special efforts. The program (which started in 1991 and is ongoing) created a racially integrated "living and learning" community in a 250-student wing of a large dormitory. It focused students on academic work (through weekly "challenge" workshops), provided an outlet for discussing the personal side of college life (through weekly rap sessions), and affirmed the students' abilities (through, for example, reminding them that their admission was a vote of confidence). The program lasted just one semester, although most students remained in the dormitory wing for the rest of their first year.

Still, it worked: it gave black students a significant academic jump start. Those in the program (about 15 percent of the entering class) got better first-year grades than black students outside the program, even after controlling for differences between these groups in the skills with which they entered college. Equally important, the program greatly reduced underperformance: black students in the program got first-year grades almost as high as those of white students in the general Michigan population who entered with comparable test scores. This result signaled the achievement of an academic climate nearly as favorable to black students as to white students. And it was achieved through a concert of simple things that enabled black students to feel racially secure.

One tactic that worked surprisingly well was the weekly rap sessions—black and white students talking to one another in an informal dormitory setting, over pizza, about the personal side of their new lives in college. Participation in these sessions reduced students' feelings of stereotype threat and improved grades. Why? Perhaps when members of one racial group hear members of another racial group express the same concerns they have, the concerns seem less racial. Students may also learn that racial and gender stereotypes are either less at play than they might have feared or don't reflect the worst-feared prejudicial intent. Talking at a personal level across group lines can thus build trust in the larger campus community. The racial segregation besetting most college campuses can block this experience, allowing mistrust to build where cross-group communication would discourage it.

Our research bears a practical message: even though the stereotypes held by the larger society may be difficult to change, it is possible to create niches in which negative stereotypes are not felt to apply. In specific classrooms, within specific programs, even in the climate of entire schools, it is possible to weaken a group's sense of being threatened by negative stereotypes, to allow its members a trust that would otherwise be difficult to sustain. Thus when schools try to decide how important black-white test-score gaps are in determining the fate of black students on their campuses, they should keep something in mind: for the greatest portion of black students—those with strong academic identities—the degree of racial trust they feel in their campus life, rather than a few ticks on a standardized test, may be the key to their success.

CONNECTIONS

In the 1920s, journalist Walter Lippmann coined the word *stereotype,* which he defined as a "picture in the mind." What does this reading suggest about the power of those "pictures in the mind"? Claude Steele writes, "In matters of race we often assume that when a situation is objectively the same for different groups, it is experienced in the same way for each group. This assumption may seem especially reasonable in the case of standardized cognitive tests." How does he challenge that assumption?

How do Steele and his colleagues use facts—particularly statistics—to define the gap between the performance of black and white students on standardized tests?

To brainstorm ideas for determining the causes? How do they test their ideas? How do they use the results to suggest remedies? Compare and contrast their methods with those of eugenicists and progressive reformers?

Commenting on the results of the intelligence test he devised, Lewis Terman, also a professor at Stanford University in the early 1900s, wrote: "The tests have told the truth." (See page 156.) How do you account for differences between his reading of the results of IQ tests and those of Steele and his colleagues?

Steele focuses on the "underachievement" of African American students who have the necessary skills and knowledge to do college work. How might his research be applied to other groups that "underachieve" in similar ways—for example, female students in science and math courses? To what extent does the notion of a "stereotype threat" apply to the way you and your classmates approach important standardized tests? Design an experiment to find out if your assumptions are correct.

Research the way at least one other social scientist views the achievement gap across the "racial divide." What questions do the studies you investigated raise? How has the scientist tried to address those questions? After you and your classmates have shared your findings, list the various solutions individuals and groups have proposed. Which do you think would do the most to bridge the gap?

---

1. This and the quotations that follow are taken from "Thin Ice: 'Stereotype Threat' and Black College Students" by Claude M. Steele. *The Atlantic Monthly*, August, 1999. Vol. 284, No. 2, pages 44-54.

Measuring Intelligences

Reading 5

Eugenicists believed that intelligence was fixed at birth and could be identified by an IQ test that measured verbal and mathematical abilities. Today few scientists still believe that intelligence is static. There is too much evidence showing that scores on an IQ test can be raised or lowered by changing a test-taker's environment. Psychologists, educators, and other researchers today also regard intelligence as far more complicated than language and mathematical skills. Howard Gardner, a psychologist who has done pioneering work on intellectual capacities, has identified the following intelligences:

Verbal-linguistic (People with this kind of intelligence enjoy writing, reading, telling stories or doing crossword puzzles.)
Logical-mathematical (Those with this kind of intelligence are interested in patterns, categories and relationships. They are drawn to strategy games and experiments.)
Bodily-kinesthetic (People with this kind of intelligence express themselves through drama, mime, dance, gesture, facial expressions, role play, and physical exercise.)
Visual-spatial (Individuals with this kind of intelligence think in images and pictures. They may be fascinated with mazes or jigsaw puzzles.)
Musical (Those who are musical are often aware of sounds others may miss. They tend to be discriminating listeners.)
Interpersonal (Individuals with this kind of intelligence are good at communicating and seem to understand others' feelings and motives.)
Intrapersonal (People with this kind intelligence are very aware of their own feelings and are often self-motivated.)
Naturalist (Individuals who are able to recognize flora and fauna, to make other consequential distinctions in the natural world, and to use this ability productively in hunting, in farming, or in the biological sciences.)

In 1999, Stefanie Weiss of the National Education Association (NEA) interviewed Gardner about his theories for the group's journal *NEA Today*. Her questions appear in italic type.

Can you give a shorthand version of your theory of multiple intelligences?

Multiple intelligences is a psychological theory about the mind.

It's a critique of the notion that there's a single intelligence which we're born with, which can't be changed, and which psychologists can measure. It's based on a lot of scientific research in fields ranging from psychology to anthropology to biology. It's not based upon test correlations, which most other intelligence theories are based on.

The claim is that there are at least eight different human intelligences. Most intelligence tests look at language or logic or both—those are just two of the intelligences. The other six are musical, spatial, bodily/kinestheic, interpersonal, intrapersonal, and naturalist.

I make two claims. The first claim is that all human beings have all of these intelligences. It's part of our species definition. The second claim is that, both because of our genetics and our environment, no two people have exactly the same profile of intelligences, not even identical twins, because their experiences are different.

This is where we shift from science to education. If we all have different kinds of minds, we have a choice. We can either ignore those differences and teach everybody the same stuff in the same way and assess everybody in the same way. Or we can say, look, people learn in different kinds of ways, and they have different intellectual strengths and weaknesses. Let's take that into account in how we teach and how we assess.

So how should teachers who believe in your theory change their approach to teaching?

. . . In my own work, I'm a proponent of teaching for understanding, which means going deeply into topics so that students can really make use of knowledge in new situations. This is very, very different from most teaching, where people memorize material and can reproduce it on demand but can't make use of it in new situations. That's what understanding entails. If you favor education for understanding the way I do, then MI [multiple intelligences] can be extremely helpful. Because when you are teaching a topic, you can approach the topic in many ways, thereby activating different intelligences. You can provide analogies and metaphors for different domains, invading different intelligences, and finally, you can present the key ideas in a number of different languages or symbol systems, again activating different intelligences.

But obviously you can't do that if you're going to spend five minutes on a topic and then move on to something. Then you're almost constrained to present it one way, which is usually verbally, and to give people a short-answer test. . . .

Can standardized tests ever hope to measure children's full intelligence?

I'm not in favor of tests that are designed to measure people's intelligence, because frankly I don't care what intelligence or intelligences people have. I care whether they can do things which we value in our culture. What good is it to know if you have an IQ of 90 or 110—or even if you can jack it up to 120 through a lot of training —if, in the end, you can't do anything.

I think our assessments ought to focus on the kinds of things we want people to understand, and they ought to give people a chance to perform their understandings. Because, at the end of the day, it doesn't matter if you have an IQ of 160 if you sit around and do nothing. What's important is whatever IQ you have or whatever profile of intelligences you have, that you can demonstrate knowledge and understanding of things that matter.

So do you think the high-stakes testing movement that we're seeing now is going to force people to abandon different approaches to teaching?

Yes. Current approaches almost inevitably push people to teach to the test, because those tests are so high-stake both for students and for teachers. Now, in principle, one could have assessments which probe understanding, and they could even be standardized. I would be much more in favor of those assessments. But those assessments would have to give people lots of choices. Because, say you're doing American history, you have to say to people, "I want you to discuss, let's say, the role of immigration in America, but you can discuss it with reference to any one of 20 different groups or 20 different issues." If, on the other hand, you require people to know all 20 different groups and all 20 issues, then obviously, they can't know very much about any one of them. It's just a very superficial, Jeopardy-style knowledge.

Now let's be clear about this: Assessment is fine. Even standardized assessment is fine, if it looks at things which are important and allows us to probe in-depth what people understand. . . .

How do you respond to those who say that MI theory is appealing, but there's no proof to back it up?

There's no short answer to that question. To begin with, it's a scientific theory, and so it needs to be evaluated on the basis of the science on which it draws. And I think it does quite well in terms of

> the scientific evidence, even the evidence that's accumulated since the theory was first propounded 20 years ago.[1]

CONNECTIONS

Howard Gardner makes two claims. The first is that all human beings have all of the intelligences he cites and that because of our genetics and our environment, no two people have "exactly the same profile of intelligences, not even identical twins, because their experiences are different." How is his view of intelligence similar to the one held by eugenicists (Chapter 5)? How does it differ? How important are the differences?

What questions does Gardner's research raise about intelligence testing? About the meaning of the word *intelligence*? Find out more about his list of multiple intelligences. To what extent do you have all eight of them? Which one best describes your style of learning?

What does Gardner mean when he says he sees multiple intelligences as a tool rather than a goal? How important is that difference to the way schools are organized? To the way teachers teach? To the way students approach their own learning?

Gardner says of intelligence tests, "At the end of the day, it doesn't matter if you have an IQ of 160 if you sit around and do nothing. What's important is whatever IQ you have or whatever profile of intelligences you have, that you can demonstrate knowledge and understanding of things that matter." Based on your study of the history of racism and the eugenics movement, what evidence can you find to support Gardner's view? To challenge that view? What do your own experiences with IQ tests add to his insights?

Gardner does not discuss the consequences of intelligence tests based solely on verbal and mathematical abilities. Find out more about those tests and how they have shaped schools in the past and the way they still affect schooling today. Share your findings with your classmates. To what extent do schools in your community still reflect the kind of categorizing and ranking that marked education in the 1900s? What do your findings suggest about the legacies of the eugenics movement?

In February of 2001, Richard C. Atkinson, the president of the University of California and founding chairman of the National Research Council's board on testing and assessment, recommended to the university's academic senate that the 10-campus system no longer require the SAT 1 for admission. Instead the university would require only standardized tests, such as SAT 2, that assess mastery of specific subjects. In April, a number of corporate leaders sent a letter to more than 70 college and university presidents urging that they place less emphasis on such tests as the SAT and ACT in admissions decisions. They argued that in their own experience character, leadership qualities, and effective communication skills matter more than test scores in determining an employee's potential. They would like colleges to apply similar criteria in their admissions procedures.

Suppose you were asked to recommend an alternative to the SAT for college admissions. What would you ask students to provide that might give college officials a better picture of their abilities? Be sure to include reasons and evidence to support your recommendations and then present them to the class.

---

1. "Meet Howard Gardner: All Kinds of Smarts," complete interview by Stefanie Weiss. *NEA Today Online. http://www.nea.org/neatoday/9903/gardner.html*

Reading 6

Many of the new scientific advances are raising tough questions for scientists, lawmakers, religious leaders, and ordinary citizens. This reading and the two that follow offer insights into current debates. In this reading, Jeff Lyon, a Pulitzer Prize-winning science writer for the *Chicago Tribune*, summarizes recent advances:

> Until recently, human cloning wasn't something most adults expected to see in their lifetimes. Even five years ago, many scientists believed it would be another 20 years or more before they figured out how to clone any species of mammal—that is, how to get a single cell from an adult animal to generate a whole new animal. But that assumption was demolished in February 1997, when British embryologist Ian Wilmut, Ph.D., announced that he and colleagues at the Roslin Institute in Edinburgh, Scotland, had successfully cloned a sheep: the now world-famous Dolly.
>
> Since then the floodgates have opened, and cattle, goats, mice, and pigs have all been cloned. Dogs haven't been cloned yet, but researchers at Texas A& M University are working on it. And now it seems it may not be long before the ultimate line is crossed.
>
> [In January 2001] Panos Zavos, Ph.D., then professor of reproductive physiology at the University of Kentucky, announced that he was leaving his position to team up with Severino Antinori, M.D., an Italian fertility specialist, to try to clone a human by 2003. Their purpose, he said, is to help infertile couples who want a genetically related child. . . .
>
> Welcome to the future, where science fiction becomes science fact and researchers and ordinary citizens alike must wrestle with a question that has profound meaning for humankind: Should scientists be allowed to pursue research that may one day enable them to shape and even create life? Or to put it another way: Is it right for scientists to assume powers that many people believe should belong only to God?
>
> Less than a decade ago, this question would have prompted an automatic answer from most people: No, it shouldn't be allowed—not that is likely to happen any time soon. But in a swift and startling turnabout, the answer to that question has become less clear, even as scientists are taking baby steps toward making such things happen. . . .

Yet, even as the likelihood of human cloning becomes more real, the science is still rudimentary. Most cloned animals die in the womb, and even those that initially seem healthy often develop fatal defects of the heart, lungs, kidney, brain, and immune system down the road. Something about cloning seems to disrupt normal gene activation in the developing fetus. This could prove catastrophic if an attempt is made to clone a human. Dr. Wilmut has said that trying it now would be "criminally irresponsible."

Nor is cloning the only sign that humans are assuming powers once relegated to the Almighty. [In September 2000] six-year-old Molly Nash of Englewood, Colorado, was given a blood transfusion that doctors hoped would help cure her of Fanconi's anemia. This rare, often fatal, hereditary disease causes the bone marrow to fail to produce blood cells and platelets. The transfused blood came from her baby brother, Adam. It had been collected from his umbilical cord at the time of his birth. Adam had been conceived in a laboratory dish with other embryos produced by his parents' eggs and sperm. He had been implanted in his mother's womb because he was disease free and because his tissue and blood type matched his sister's—in other words, so he could be her donor. The other embryos were discarded. Cord blood is rich in stem cells, the mother cells found in various organisms that generate the functional cells of those organs. It was hoped that Adam's stem cells would generate functioning bone marrow and a healthy new blood supply for Molly.

The procedure seems to have worked. Tests done in January [2001] found that almost all of Molly's bone marrow came from Adam. "While we will continue to monitor Molly, especially over this first critical year, her prognosis looks great," said John Wager, M.D., a transplant specialist at the University of Minnesota Medical School, who performed the transfusion. The Nashes did not doubt they had done the right thing. "You could say it was an added benefit to have Adam be the right bone-marrow type, which would not hurt him in the least and would save Molly's life," Lisa, their mother, said in September. "We didn't have to think twice about it." But some ethicists were concerned. Would children now be bred for their biological usefulness?

Stem cells, meanwhile, are the focus of another scientific endeavor that rivals cloning in its potential to bestow Godlike powers on human beings. Researchers hope someday to be able to direct a person's stem cells to grow new organs and tissues for that person in a lab. The cells could be told to grow a liver for someone who needs a transplant, for example, or brain cells for someone with Alzheimer's disease. And because the cells would contain the person's own DNA, there would be no problem with tissue rejection.

Advances in genetic engineering and gene therapy are also transforming the nature of life and the way we live. Researchers have already created genetically altered seeds and grains designed to produce hardier plants and bigger harvests—and American consumers are already eating some of this altered produce without knowing it. And despite a tragic setback in September 1999, when 18-year-old Jesse Gelsinger of Tucson, Arizona, died during a gene-therapy experiment at the University of Pennsylvania, research is also moving forward in developing safer, more effective ways to deliver healthy new genes into a patient's cells.

Thanks to the Human Genome Project, the ongoing effort to codify and learn the function of the . . . genes that make up the instruction manual for the human body, researchers are also zeroing in on which genes cause and can cure various diseases. In a few years it may be possible for people to go to a doctor's office, and in the time it takes to read this article, get a full lab report detailing their genetic predisposition to various diseases. If the report noted a susceptibility to lung cancer, for example, they would then be counseled not to smoke. In the not-too-distant future, scientists could also have the power to design smarter, more attractive, and athletic offspring by tinkering with a child's genetic makeup before or after birth. Such powers would enable them to change the course of human evolution, and do it in a matter of generations.

And then there is the ultimate quest: to create life itself. In 1953 researchers at the University of Chicago mixed methane, ammonia, hydrogen, and water—the ingredients of the so-called "primordial soup" that existed on the young earth-and passed an electric current through it to simulate lightning. To their amazement, they found traces of amino acids—the chemical building blocks of life—in the residue. Now a team of scientists headed by a brilliant maverick named J. Craig Venter, Ph.D., director of the Institute for Genomic Research in Rockville, Maryland, is conducting another experiment.

Working with a harmless species of bacteria called Mycoplasma genitalium that has only 517 genes—the fewest of any known organism—Dr. Venter and his colleagues disrupted the microbe's genes one by one to see which it needed to stay alive. The next task, they wrote, is to narrow down that number as a "first step" toward "engineering" a cell with "a minimum genome" in the lab: in other words, manufacturing a living microbe.

That's as far as Dr. Venter has taken the research. The question is whether anyone should take it any further. In the issue of Science containing his paper, a panel of bioethicists—thinkers who specialize in weighing the thorny issues raised by modern medicine and biology—addressed this point at his request. They . . . gave it a conditional thumbs-up. The prospect of humans creating a life form "does not violate any fundamental moral precepts," the authors wrote. But they did raise questions they felt needed to be considered, such as whether the new technology would "be used for the benefit of all" and the possibility that it could be misused to create new biological weapons. [1]

CONNECTIONS

What evidence does Jeff Lyon offer of "science fiction" becoming "science fact"? What evidence can you add based on your study of the history of racism and the eugenics movement?

What do efforts to clone animals and ultimately human beings suggest about the power of ideas? About the way an idea that seems repulsive at first becomes more and more attractive? What aspects of the history of the eugenics movement may offer scientists, politicians, and ordinary citizens useful insights as they consider the possibilities of cloning?

Invite one or more guest speakers to the class to address the implications of genetic testing. You might ask a researcher in biotechnology, a physician, or someone knowledgeable about the implications of genetic testing for people with disabilities or inheritied illnesses to address the class.

Find out more about genetic testing by researching one or more of the following diseases, disabilities, or conditions. Or you may prefer to study one of your own choosing.

Diseases:	**Disabilities:**	**Conditions:**
Tay Sachs	Blindness	Dwarfism
Cystic Fibrosis	Down Syndrome	Baldness
Muscular Dystrophy	Spina Bifida	Cleft Palate
Breast Cancer	Fragile X Syndrome	
Sickle Cell Anemia		

As you gather information, look for answers to the following questions:

- ·What is the genetic basis of the condition? (Single gene, polygenetic, etc.) To what degree does prenatal development, diet, and the environment influence its development?
- ·What do the existing technologies reveal about the condition? What remains to be learned?
- ·What are the critical questions for us as citizens? These questions may relate to individual choices or public policy.

Present your research to the class. What concerns emerge as you listen to other reports and compare them to your own? What are the implications of those concerns?

Collect recent articles on genetic research. Read at least five of the articles and list the claims and cautions the authors make about genes. Compare your findings with those of your classmates. How do you account for similarities among the articles? To what extent do they support concerns about "genetic determinism"? To what extent do they challenge that idea?

---

1. "Playing God: Has Science Gone too Far?" by Jeff Lyon. *Family Circle*, July 10, 2001, pp. 56, 58.

Reading 7

Arthur Caplan, a bioethicist, says of Craig Venter's efforts to manufacture a living microbe (Reading 6), "A couple of years ago I'd have opposed this experiment. I think society is becoming used to genetic tinkering." That is exactly what worries Laura Hershey, a Colorado consultant who served on the Denver Commission for People with Disabilities. She is among the disability-rights activists who are "becoming increasingly alarmed about the economic and political issues arising from the rapidly advancing field of genetic research." In 1999, she wrote in part:

> The application of genetic knowledge to the repair of damaged genes, for the purpose of treating certain illnesses, may offer welcome benefits to some people with disabilities. But genetic research is likely to be put to other, more insidious, uses such as denying health insurance, even jobs, to people whose genes predispose them to medical problems. Another threat is the implementation of eugenic policies to "weed out" certain types of people from the population. Thus, along with the much-heralded scientific advances offered by genetic research, disability activists nervously witness a resurgence of eugenic thinking.
>
> **Genetic Screening Against Disability**
>
> Using ultrasound and abortion to select a child's sex is regarded as unacceptable to most people. Using genetic testing to eradicate characteristics such as homosexuality is still a new concept, but is likely to cause a great deal of controversy. Yet the media and the public seem to accept, almost without question, the idea of screening for genetic anomalies that cause disabilities and then using that information to eliminate certain conditions, by eliminating their carriers before birth.
>
> Scientists and journalists may consider genetic screening against disability a wise public health strategy. But the progressive disability community sees the dangers inherent in targeting genetic research toward efforts to do away with disability. . . .
>
> Many people assume that people with disabilities would want to spare future generations from the difficulties we had to endure. But this assumption relies on another assumption, that our disabilities are inherently problematic. The disability-rights movement disputes that

idea. Rather than blaming our physical or mental disabilities themselves, we see our problems as rooted in social, physical, economic and political barriers. Attempting systematically to wipe out disabilities is the wrong solution. Instead, society should commit itself to removal of these barriers, and to full equality for people with disabilities.

Still, why would disabled adults object to genetic practices which do not directly affect us? At first glance, genetic screening seems to target only potential people with disabilities—either fetuses diagnosed with genetic anomalies, or those not yet conceived, but at risk of such anomalies. But in fact, the mindset that advocates the widespread, even routine use of screening also promotes efforts to "prevent disability"—not by reducing occupational hazards and violence, nor by improving health care or environmental conditions; but by deterring the births of children who may have disabilities.

Genetic counseling, prenatal testing, and selective abortions arise from—and reinforce—the erroneous and dangerous belief that people with disabilities are a problem. As our society struggles with the allocation of health care resources, we overlook the vast amounts of money which are consumed by corporate bureaucracies and private profits. People with disabilities are scapegoated for needing and using expensive medical services and ongoing supports. . . .

As an example, witness the recent remarks of Dr. Bob Edwards, world-renowned embryologist and creator of Britain's first test-tube baby. Speaking at an international fertility conference, Edwards said the increasing availability of prenatal screening for genetic disease gave parents a moral responsibility not to give birth to disabled children. Edwards celebrated a new age in which every child would be genetically acceptable. "Soon," he pronounced, "it will be a sin of parents to have a child that carries the heavy burden of genetic disease. We are entering a world where we have to consider the quality of our children.". . .

Not Model Citizens

Since virtually the beginning of the disability-rights movement, activists have critiqued "the medical model." This model viewed people with disabilities—our bodies, our social identity, our private histories—as pathology. The medical model viewed people with disabilities as afflicted, ill, aberrant, burdened patients to be cured, or at least rehabilitated.

We refuted the mastery of the physician, and challenged the

built-world around us to change, to adapt to our nonstandard specifications. The disability-rights movement insists on accessibility and accommodations, not as benevolent gestures toward the "less fortunate" but as the civil rights of a large political minority.

Increasingly, another ideology is evolving from the medical model. The field of public health has gained prominence in recent years, spawning new, perhaps equally coercive beliefs about disability.

Under the public health model, one person's health or illness becomes a societal responsibility. Health equals good citizenship, whereas illness is expensive, disruptive, and (with genetic intervention) can be preventable.

For all its oppressiveness, the old medical model did claim as its primary concern the well-being of the patient herself. Its definitions and prescriptions could be profoundly misguided, but they were made in the name of serving the disabled person's needs. In contrast, the public health model aims to serve the dominant (nondisabled) majority, by cutting costs associated with disability. As disability-rights advocate, author, and psychologist Carol Gill points out, the idea of "promoting wellness" sounds benign—but in practice, it can mean that "disenfranchised people suffer."

A Place at the Research Table

This isn't just a matter of good science being used for bad purposes. Disability activists question the research itself; we deserve and demand an opportunity to give input into the directions taken by the Human Genome Project and other research endeavors. This means questioning the presumption of total scientific objectivity.[1]

CONNECTIONS

What do Arthur Caplan's comments suggest about the way a society becomes used to an idea? To what extent does Laura Hershey challenge that notion? What does history teach us about the way ideas take root in a society? Do changes happen all at once? Or are they made little by little, step by step?

Hershey critiques "the medical model" for viewing people with disabilities. How does she characterize that model? How does she contrast it with what she calls "the public health model"? Research both models. What do your findings suggest about the consequences of the way we define one another? About the power of ideas to divide as well as unite people? Why might those divisions encourage separation, conflict, and even violence?

What does Hershey see as the relationship between science and society? What arguments does she use to suggest the way that relationship determines the way people define their universe of obligation? Their ideas about "good citizenship"? In reflecting on your own reading and experience, what events, speeches, or arguments would you add in support of her point of view? What events, speeches, or letters might be used to question her point of view?

Chapter 1 featured an episode from *The Twilight Zone* entitled "The Eye of the Beholder." It offered a provocative answer to the question "What do you do with a difference?" How does Laura Hershey answer that same question? What does she add to our understanding of such words as *normal* and *healthy*? To what extent is "health" in "the eye of the beholder"?

Medicine is generally viewed as a healing profession and science as a body of knowledge that advances society. What was being "healed" in the society featured in "The Eye of the Beholder"? How was society being "advanced"? What did the episode suggest about the way physicians and scientists promote the values of their society? What did it suggest about the way the values of the larger society influence their work? What does Hershey add to your understanding of those questions? Of the importance of our answers to those questions?

Find out more about the disability-rights movement. When did it begin? To what extent is it an attempt to learn from history? To undo the legacies of that history? What new questions does it raise? How would you go about finding answers to those questions?

---

1. "Disability Rights Activisits Warn of Eugenics" by Laura Hershey. *Resist,* September, 1999. Copyright Resist, Inc.

Raising Moral Questions

Reading 8

As Laura Hershey's comments reveal, genetic research raises tough questions: What does it mean to be human? What is normal? When does life begin? Jeff Lyon, a science writer for the *Chicago Tribune*, summarizes recent discussions focusing on those questions:

> Different people simply have different beliefs about how life came to exist and where humans fit in the grand design. "I see life as a process of chemistry," says Norman Pace, Ph.D., a professor of nuclear, cellular and developmental biology at the University of Colorado who is involved in his own quest to isolate the minimal components of life in the lab. "I see life as chemicals talking to one another in sophisticated ways developed through natural selection. Much of it we don't yet understand, but that doesn't mean it's a spiritual matter. These spiritual matters are human inventions."
>
> Even if God exists, say others, we can't call these pursuits "playing God" because they don't reflect how God operates. "In nature, chance determines things," says R. Alta Charo, J.D., professor of law and medical ethics at the University of Wisconsin Law School. "I believe that the essence of God is to let the odds play out." In contrast, she says, "It is the essential attribute of being human to make choices, to exercise control, to have dominion over the natural world." She sees these quests as "completely consistent with what it means to be human on this planet. I believe knowledge is an intrinsic good and that until it is shown to cause harm, it should be encouraged. I believe we should have eaten the apple."
>
> Not everyone shares these views. Lori Andrews, Ph.D., a professor of law at the Chicago-Kent College of Law and a legal specialist in new reproductive technologies, thinks ethicists have become too accepting of a whole laundry list of unsettling scientific quests. "It's like we've become deadened to the ethical dimensions of this," she says. "We're viewing biology as playing with Tinker Toys. There seems to be less resistance to the whole idea of tampering with life.". . .
>
> Richard Hays, former assistant political director of the Sierra Club, finds the lack of loud public debate about [new] technologies "chilling" and holds bioethicists partly to blame. "Many of these academics have become almost apologists for genetic engineering and cloning," says Hays, now executive director of the Exploratory

Initiative on the New Human Genetic Technologies, a network of professionals and activists interested in stimulating that debate. "You rarely find a bioethicist who thinks there's anything fundamentally wrong with these technologies. In Europe it's very different, because they had the Nazi Holocaust. But here we have consumer-driven markets.

Not all bioethicists fit this mold, of course. Leon Kass, M.D., Ph.D., the Addie Clark Harding professor in the Committee on Social Thought at the University of Chicago, is one who doesn't. It worries him, he says, "that the scientists' view of what they're doing could rapidly become the public's view, and that kind of shrunken understanding of what life is—that it's nothing but chemicals—could spread even further in the culture than it already has. It seems to support the materialist view of life—which, even though I'm a trained scientist, I regard as false and inadequate."

Dr. Kass argues further that making a microbe in a lab is not really creating life. "It's a gross exaggeration. It's like reproducing a Mozart symphony. You haven't written the score; you are merely recopying it. I'm bothered that we are coming under the illusion that because we know how to reproduce a few things, we are absolutely in charge. It's a form of hubris and folly." Besides, he says, even if a scientist could create a human from scratch, "would he really be the author or just the instrument of God's handiwork?"

Lisa Sowle Cahill, Ph.D., J. Donald Monan chair of theology at Boston College and former president of the Catholic Theological Society of America, wonders about this, too. "The Bible says we are created in the image of God and God is the Creator," she says. "Does that mean only God creates? Or does it mean that because we are made in God's image we share that ability? If so, who is to say which of our efforts do and don't cross the line? Are we playing God when we wipe out smallpox or cure cancer? Why is it wrong to put a jellyfish gene in a monkey?" It makes us uncomfortable for many reasons, she says, "but defining why it is wrong is more difficult—for me, anyway."

Like many religions, the Catholic Church "doesn't have a final position on a lot of these questions," says Dr. Cahill. "It cautiously welcomes new genetic therapies, but it is concerned about protecting human life and has ruled out research using human embryos. Other things are not settled."

But religion can guide and prod people to think in ways they otherwise might not. "It is the nature of religion to be conservative," says Harold S. Kushner, Rabbi Laureate of Temple Israel in Natick, Massachusetts, and author of the forthcoming *Living a Life That Matters: Resolving the Conflict between Conscience and Success.* "Religion says, 'Wait a minute, there are time-tested values here which we should be very slow to disregard.' I'd hope our experience with polluted air and toxic and nuclear waste would have taught us not to go where we can just because we can. I'd hope for a self-imposed moratorium on doing what's possible until we figure out whether we really want to do it."

"In vitro fertilization is wonderful," says Rabbi Kushner. "DNA repair is good. My wife and I had a son who died of a genetic disease, and the idea of fixing what's missing and giving an innocent child life is exciting. But it is one thing to repair, and another to let parents make sure that they have perfect children. My concern is we will lose the knack of loving children who are less than perfect. And my concern with cloning is less ambivalent. I mind very much if we clone people. The whole idea of God's plan for humanity, which calls for people to have children and die, means that one generation, scarred and wearied by its experience, gives way to another that's born fresh and innocent and full of promise. Once you start fooling with that, I think you undermine what God has in mind for the human race. As for creating life artificially, there is something special about humans being created out of an act of love, not chemistry."

Dr. Kass agrees. There is a difference between using the new technologies to cure disease and "using them to engineer so-called improvements," he says. "As a species we don't have the wisdom to know what an improvement would be. The better path is caution and humility before these awesome powers we may never fully understand." Indeed, says Rabbi Kushner, "A scientist ought to stand in awe of the things modern science can do and realize that he has seen the face of God, he hasn't become God."

One thing is clear. These technologies are here to stay, and it's up to all of us to decide what to do with them. "We want to support the most creative and compassionate science possible," says Laurie Zoloth, Ph.D., head of the Jewish Studies department at San Francisco State University. "The bold scientific approach allowed Pasteur and Salk to take leaps that advanced the cause of humankind. But the human capacity for error is enormous. And the human capacity for terrible moral choices is also great. We live in a society in which some 44 million people have too little access to health care. And now we're developing technologies that may give enormous life-shaping power to people who have the money to control it. So there is a lot to be cautious about."

Hays is more blunt. "What's at stake is our common human future. Genetic modification could lead to the creation of separate genetic castes and social division beyond anything in history. There's no reason to go down this road. We need to summon the maturity to use our technology in ways that affirm rather than degrade humanity. We have to decide which uses we approve of and which we oppose."

The only way we can do that, says Dr. Zoloth, is through an "enormous national conversation. All we have is the ability to keep talking and raising fears and hopes and encouraging scientists to stop and reflect." History shows we can achieve great things if we keep talking. "When we wanted to think about race, we had a transformative national conversation. The civil-rights movement was America at its best. The Vietnam War sparked such a conversation. Now we need to have one about genetics. This is exactly the moment when we must decide who controls this technology and on behalf of whom. The need cannot be overestimated. This is far too important to leave in the hands of market forces alone."[1]

Sociologist Barbara Katz Rothman suggests why many people are reluctant to enter into such conversations:

I'm a sociology professor; I get paid to read. I can afford to take a couple of years and read in genetics and bioethics. Most people probably cannot do that; they have other things to do. But the conclusions that I have come to, from all of that technical reading in genetics and in bioethics, is that you don't need the technical understanding to make the moral judgments.

A group of sociologists in Scotland came to the same conclusion. They ran focus groups of lay people on ethical issues in genetics. They concluded, "Technical competence was neither relevant or important to the majority of participants in our study: they discussed issues without need to display technical competence. When the technical issues were mentioned, the accuracy of the knowledge was irrelevant to the point being made." They gave an example of a group discussion in a working-class area of Edinburgh: "They are going a little too far. If they want to go and investigate the DNA system and found out that OK somebody's gay because there is a little slip-up in the XY hormone, we can do an injection and fix that, or a kid's going to be born mongoloid, rather than abort we may be able to find a way that we can actually sort the gene out. We are getting to the part with genetic engineering if somebody is going to get a deformed child then they just get rid of it and say 'right the next one you produce will be.'"

This person is completely wrong on every technical point going. XY isn't a hormone; mongoloid isn't the current word and it's not a "gene" to be "sorted out." And so what? The question that the person is raising is about drawing moral lines, about drawing lines and going too far. Again, you or I may or may not agree with him, just as we may or may not agree with far more sophisticated language the theologians used. But moral authority does not rest on technical authority: the concerns that are being raised, including the concerns that you personally may feel, are in and of themselves worth discussing.

Genetics, as a science, as a practice and as an ideology, is offering us a great deal. But we have to decide if we want what it has to offer. Those decisions are not technical matters. The technology of it all is overwhelming. Keep bandying about terms like "alleles," "RFLPS," "clines," "22Qlocus," and most of us are left in the dust. Promise a cure for cancer, and end to human suffering, and it's hard to argue. Troy Duster puts it, "Technical complexities of vanguard research in molecular biology and the promises of success incline us to go limp before such scientific know-how."

We cannot afford to go limp. We'll be carried off to places we might very well choose not to go.[2]

CONNECTIONS

In Chapter 7, physicist Leon Lederman was quoted as saying that scientific knowledge is "not good or evil; it is enabling. Modern science, however abstract, is never safe. It can be used to raise mankind to new heights or literally to destroy the planet . . . . We give you a powerful engine. You steer the ship." What does this reading suggest about the difficulties in "steering the ship" in this age of genetic engineering? About the role of a citizen in a democracy in the 21st century?

A number of individuals quoted in this reading speak of the need for "loud public debate." What might such a debate look like? Where might it occur?

Whose voices should be heard? How might those individuals and groups advocate for their points of view?

According to an old saying, a little knowledge is a dangerous thing, but show me the person with so much knowledge that he or she is out of danger. What is the moral or lesson of the saying? To what extent does this reading support that lesson? To what extent does it challenge the lesson?

Rothman believes that people find the language of bioethics and genetics too complicated for the average person. She writes, "Mystification is a political tool: making something complicated is a way of disempowering people." In what sense is technical language "mystifying"? Why does Rothman believe that mystification is disempowering? Why does she believe that technical language is unnecessary to the central issues in current debates?

How are the issues raised in debates about genetic engineering and cloning similar to the public health issues eugenicists raised in the early 1900s? What differences seem most striking?

In reflecting on the debate over stem-cell research, Kenneth L. Woodard writes:

> In any political debate burdened by strong ethical differences, the first casualty is usually language itself. So it is with the ethical issues surrounding stem-cell research—specifically the question of whether days-old human embryos should be destroyed on the promise they offer of therapeutic answers to Parkinson's and other degenerative diseases. The words we choose to frame our arguments reveal the moral universe we inhabit. Those tiny flecks frozen in tanks of liquid nitrogen—what exactly are they? To the secular eyes of The New York Times editorial page, for example, they are "just clumps of microscopic cells" and thus of no intrinsic moral worth. On the other hand, what the Vatican sees is the moral equivalent of a fully developed "person" and therefore worthy of social respect and legal protection. Most everyone else sees something in between.[3]

Why does it matter what words we choose to frame our arguments? Gather information about the debate over stem-cell research. What words does each side use to express its hopes and fears? What might history add to the discussion? How might the lessons of the past help all sides in the debate find common ground?

The new technologies raise important issues about what it means to be human. Working in small groups, create a chart showing where each of the scientists, theologians, and other thinkers quoted in this reading stand in regard to genetic research. Which oppose any limitations on genetic research? Which favor no genetic research? Which fall somewhere in the middle? Analyze your chart. What do the various groups have in common? Whose position is closest to your own?

Working alone or with a partner, find out how at least one other theologian, philosopher, or other thinker views genetic research. Compare and contrast his or her views with those outlined on your chart and those of individuals your classmates researched. What do the various answers suggest about what it means to be human in the 21st century? To be a "good citizen" in this new age?

Find out more about the work of bioethicists. What role do they play in scientific inquiry? What do they add to the process? What are the risks in their work? If possible, invite a bioethicist to speak to the class. Meet in small groups to formulate a list of questions to ask about his or her work. Try to keep your questions open-ended so that you can learn how he or she thinks about an issue, assesses a risk, or judges an outcome.

---

1. "Playing God: Has Science Gone too Far?" by Jeff Lyon. *Family Circle,* July 10, 2001, pp. 60, 62, 63.
2. Barbara Katz Rothman, *The Book of Life: A Personal and Ethical Guide to Race, Normality, and the Implications of the Human Genome Project*. Beacon Press, 1998, 2001, pp. 38-39.
3. "A Question of Life or Death" by Kenneth L. Woodward. *Newsweek,* July 9, 2001, p. 31.

The Power of History

Reading 9

A number of important questions have guided your study of the history of racism and the eugenics movement: What do we do with a difference? What does it mean to be human? How do we understand human differences? How do we as individuals and as citizens define our universe of obligation? Eugenicists thought that they had clear answers to these questions. They promoted their vision for the nation as scientific and rational even though more often than not their vision was rooted in myth and dogma. Wherever that vision was translated into public policy, the consequences were alarming and too often deadly.

This book has shown that every event, every movement in history, has consequences. It touches not only those who experienced it but also their children and their children's children. Our identity is shaped, at least in part, by our history. How do we remember this history? How can we prevent it from happening again? A number of years ago, a principal answered these questions by sending the following letter to teachers on the first day of the school year:

> Dear Teacher:
>
> I am a survivor of a concentration camp. My eyes saw what no man should witness:
>
> -Gas chambers built by learned engineers.
> -Children poisoned by educated physicians.
> -Infants killed by trained nurses.
> -Women and babies shot and burned by high school and college graduates.
>
> So I am suspicious of education.
>
> My request is: Help your students become human. Your efforts must never produce learned monsters, skilled psychopaths, educated Eichmanns.
>
> Reading, writing, arithmetic are important only if they serve to make our children more humane.[1]

After reflecting on the lessons of history and his own experiences as an artist in the United States, Jos. A. Smith, a children's book illustrator, wrote a brief essay entitled "Your Kind."

> The greatest threat we pose to each other is a fruit of our sublime ability to generalize.
>
> The capacity to manipulate symbols–the root of our talent to

learn and theorize—is also the source of our art.

And ah, see how creatively we use it!

After all, let me transform you into an abstraction and I have permission to deprive you of your basic rights, your freedoms, even (and this is really only another small step) your life.

As long as I see you as a person, I'm lost. If you remain someone who has needs, who laughs and cries, and who feels pleasure or pain, I see a real person who might stop to pet a dog or marvel at a poem. It's too easy to care for you. I might even be tempted to share what I have with you.

Let me turn you into a symbol and you are nothing but a label. I push you back to an emotional distance beyond my power to focus. The details that make you real disappear. You blend into a faceless group I can call "Your Kind."

Thank God I'm not one of "Your Kind."

As long as we divide people into "Us" and "Them," let's not pretend to be surprised when evil smiles back at us from the mirror.[2]

CONNECTIONS

How does the principal seem to define the word *human*? Compare and contrast his definition with others quoted in this book. What similarities do you notice? What differences seem most striking? Which definition is closest to your own?

What importance does Smith place on symbols? Why does he believe that they must be manipulated with care? How does his warning relate to the history of racism as well as the history of the eugenics movement? How does it relate to current events?

In the introduction to this chapter, German historian Detlev J. K. Peukert was quoted as saying, "The shadowy figures that look out at us from the tarnished mirror of history are—in the final analysis—ourselves." How does Smith underscore the importance of that idea? How do you think he would answer the central question of Chapter 1: What do we do with a difference? How would you answer it now that you have studied the history of racism and the eugenics movement?

One way a community preserves memory and confronts its history is through monuments that honor its heroes, mourn its victims, or commemorate its tragedies. What do you think would be an appropriate way of remembering the history of racism and the eugenics movement? What would you want visitors to remember? What would you want them never to forget?

Design a monument to some aspect of the history of racism and/or the eugenics movement. For ideas, you may want to explore the monuments and memorials section at facinghistory.org. Share your creation with your classmates by explaining the purpose of your memorial and what you hope your intended audience will learn from it.

---

1. Quoted in *Teacher and Child* by Haim Ginott. Macmillan, 1972, p. 317.
2. "Your Kind" by Jos. A. Smith in *Tikvah: Children's Book Creators Reflect on Human Rights*. Edited by Norman D. Stevens and Elie Wiesel. SeaStar Books, 1999, p. 80.

For Further Investigation

This section of *Race and Membership in American History: The Eugenics Movement* is divided into two parts. Part 1 identifies general books and websites that provide useful information on the history of racism, eugenics, and/or American citizenship. Part 2 provides for each chapter a list of relevant literature, reference materials, websites, and videos. An annotated version of For Further Investigation and an explanation of Facing History's borrowing policy may be found at *www.facinghistory.org.*

Part 1: General Resources

Books

Bieder, Robert Eugene. *Science Encounters the Indian, 1820-1880: The Early Years of American Ethnology.* University of Oklahoma Press, 1995.

Chase, Allan. *The Legacy of Malthus: The Social Costs of the New Scientific Racism.* Knopf, 1977.

Gould, Stephen Jay. *The Mismeasure of Man.* W.W. Norton, 1996. (Revised and expanded edition.)

Jacobson, Matthew Frye. *Whiteness of a Different Color.* Harvard University Press, 1999.

Kevles, Daniel J. *In the Name of Eugenics: Genetics and the Uses of Human Heredity.* Harvard University Press, 1995.

Marks, Jonathan. *Human Biodiversity: Genes, Race, and History.* Aldine De Gruyter, 1995.

Proctor, Robert N. *Racial Hygiene: Medicine Under the Nazis.* Harvard University Press, 1989.

Selden, Stephen. *Inheriting Shame: The Story of Eugenics and Racism in America.* Teachers College Press, 1999.

Smith, Rogers M. *Civic Ideals: Conflicting Visions Of Citizenship In U.S. History.* Yale University Press, 1997.

Takaki, Ronald T. *A Different Mirror: A History of Multicultural America.* Little, Brown, & Co., 1994.

Torres, Rodolfo D., Louis F. Mirón, and Jonathan Xavier Inda, ed. *Race, Identity, and Citizenship: A Reader.* Blackwell Publishers, 1999.

Websites

American Eugenics Society Scrapbook
http://www.amphilsoc.org/library/guides/eugenics.htm

Brief Timeline of American Literature and Events
http://www.gonzaga.edu/faculty/campbell/enl311/timefram.html

Center for Immigration Studies *http://www.cis.org/*

Image Archives on the American Eugenics Movement
http://www.eugenicsarchive.org/eugenics

Scope Note on Eugenics
http://www.georgetown.edu/research/nrcbl/scopenotes/sn28.html

Part 2: Resources by Chapter

Chapter 1: Science Fictions and Social Realities

Facing History Resources

Facing History and Ourselves: Holocaust and Human Behavior—Chapter 1 and Reading 1 of Chapter 2, "Harrison Bergeron," a short story by Kurt Vonnegut.

Videos

Eye of the Beholder. Movies Unlimited (22 min.) See Reading 1, pp 4-6.
Facing Evil. Film for the Humanities (60 min.)
Jefferson's Blood. PBS (90 min.)
Masterpiece Society. Paramount (46 min.) See Reading 8, pp. 31-32.

Websites

Asian American Studies Resource Guide *http://www.usc.edu/isd/archives/ethnicstudies/asian*
Documenting the American South (DAS) *http://docsouth.unc.edu/index.html*
Writing Black: *http://www.keele.ac.uk/depts/as/Literature/amlit.black.html*

Novels/Memoirs & Autobiographies

Chin, Frank. *Donald Duk: A Novel.* Coffeehouse Press, 1991.
Cisneros, Sandra. *The House on Mango Street.* Vintage Books, 1991. (Reissue edition.)
Gates, Henry Louis, Jr. *Thirteen Ways of Looking at a Black Man.* Vintage Books, 1998.
McBride, James. *The Color of Water: A Black Man's Tribute to His White Mother.* Riverhead Books, 1997.
O'Hearn, Claudine C., ed. *Half and Half: Writers on Growing Up Biracial and Bicultural.* Pantheon Books, 1998.
Riley, Patricia, ed. *Growing Up Native American.* Morrow, 1993.
Singer, Bennett L., ed. *Growing Up Gay/Growing up Lesbian.* New Press, 1994. (Reprint edition.)

Reference

Atkins, Dawn. *Looking Queer: Body Image and Identity in Lesbian, Bisexual, Gay and Transgender Communities.* Harrington Park, 1998.
Correspondents of the *New York Times. How Race Is Lived in America: Pulling Together, Pulling Apart.* Times Books, 2001.
Fries, Kenny, ed. *Staring Back: The Disability Experience from the Inside Out.* Plume, 1997.
Minow, Martha. *Making all the Difference: Inclusion, Exclusion, and American Law.* Cornell University Press, 1991.
Nelkin, Dorothy and M. Susan Lindee. *The DNA Mystique: The Gene as a Cultural Icon.* W.H. Freeman & Co., 1996.
Spencer, Rainier. *Spurious Issues: Race and Multiracial Identity Politics in the United States.* Westview Press, 1999.

Walker, Rebecca. *Black, White, and Jewish: Autobiography of a Shifting Self.* Riverhead Books, 2000.

Wolf, Naomi. *The Beauty Myth: How Images of Beauty Are Used Against Women.* Anchor, 1992.

Chapter 2: Race, Democracy, and Citizenship

Facing History Resources

Facing History and Ourselves: Holocaust and Human Behavior—Chapter 2

Videos

Africans in America. PBS (four 90-min. episodes)
In The White Man's Image. PBS (58 min.)
Jefferson's Blood. PBS (90 min.)

Novels/Memoirs & Autobiographies

Douglass, Frederick. *Narrative of the Life of Frederick Douglass, an American Slave,* ed. by John W. Blassingame et. al. Yale University Press, 2001. Also available online at *www.ipl.org.*

Equiano, Olaudah. *Interesting Narrative,* ed. by Robert J. Allison. Bedford Books, 1995.

Walker, Cheryl. *Indian Nation: Native American Literature and Nineteenth Century Nationalism*. Duke, 1997

References

Bieder, Robert Eugene. *Science Encounters the Indian, 1820-1880: The Early Years of American Ethnology.* University of Oklahoma Press, 1995.

Frederickson, George M. *The Black Image in the White Mind: The Debate on Afro-American Character and Destiny.* Harper & Row, 1971.

Gossett, Thomas F. *Race: The History of an Idea in America.* Oxford University Press, 1997. (2nd ed.)

Graves, Joseph L., Jr. *The Emperor's New Clothes: Biological Theories of Race at the Millenium*. Rutgers, 2001.

Harding, Sandra, ed. *The 'Racial' Economy of Science: Toward a Democratic Future*. Indiana University Press, 1993.

Higginbotham, A. Leon. *In The Matter Of Color: The Colonial Period.* American Philological Association, 1978.

LaCapra, Dominic, ed. *The Bounds of Race: Perspectives on Hegemony and Resistance*. Cornell University Press, 1991.

Smith, Rogers M. *Civic Ideals: Conflicting Visions Of Citizenship In U.S. History.* Yale University Press, 1997.

Stanton, William. *The Leopard's Spots: Scientific Attitudes Toward Race in America, 1815-1859.* University of Chicago Press, 1960.

Takaki, Ronald. *Iron Cages: Race and Culture in 19th Century America.* Oxford University Press, 2000.

Chapter 3: Evolution, "Progress," and Eugenics

Facing History Resources

Facing History and Ourselves: Holocaust and Human Behavior—Chapter 2

Videos

The First Measured Century. PBS (two 90-min. episodes)
Homo Sapiens 1900. First Run Icarus (88 min.)
In Search Of Ourselves. PBS (120 min.)

Novels/Memoirs & Autobiographies

Bellamy, Edward. *Looking Backward.* Dover Publications, 1996.
Chestnutt, Charles Waddell. *The Conjure Woman.* Originally published 1899. Available as a free download from the Internet Public Library *(www.ipl.org).*
Chopin, Kate. "Desiree's Baby." collected in *The Awakening* and *Selected Short Stories of Kate Chopin.* Signet/New American Library, 1995.
Ellison, Ralph. *Invisible Man,* 2nd edition. Vintage Books, 1995.
Faulkner, William. *Light in August: The Corrected Text.* Vintage Books, 1991. (Reissue edition.)

Reference

Baker, Lee D. *From Savage to Negro: Anthropology and the Construction of Race, 1896-1954.* University of California Press, 1998.
Bannister, Robert C. *Social Darwinism: Science And Myth In Anglo-American Social Thought.* Temple University Press, 1988.
Hofstadter, Richard. *Social Darwinism in American Thought.* Beacon Press, 1992. (Reprint edition.)
Rafter, Nicole Hahn. *White Trash: The Eugenic Family Studies, 1877-1919.* Northeastern University Press, 1988.

Chapter 4: In an Age of "Progress"

Facing History Resources

Facing History and Ourselves: Holocaust and Human Behavior—Chapters 2 and 3

Videos

The Bontoc Eulogy. Cinema Guild (57 min.)
The First Measured Century. PBS (two 90-min. episodes)
Homo Sapiens 1900. First Run Icarus (88 min.)
New York, Episode 4 "The Power and the People, 1898-1918," Warner Home Video (120 min.)
World on Display. New Deal Film (53 min.)

Websites

American Memory: Historical Collections for the National Digital Library *http://memory.loc.gov/*

Interactive Guide to the World's Columbian Exposition
http://users.vnet.net/schulman/Columbian/columbian.html
Louisiana Purchase Exposition
http://www.boondocksnet.com/expos/louisiana.html
Smithsonian National Museum of American History, "Between a Rock and a Hard Place: A History of Sweatshops, 1820-Present"
http://americanhistory.si.edu/ve/index.html
The World's Columbian Exposition of 1893 Collection
http://cpl.lib.uic.edu/001hwlc/speworldexp.html

Novels/Memoirs & Autobiographies

Addams, Jane. *Twenty Years at Hull-House: with Autobiographical Notes*. Signet, 1999.
Crane, Stephen. *Maggie, a Girl of the Streets.* Fawcett, 1995. (Reissue edition.)
Dreiser, Theodore. *Sister Carrie*. Signet, 2000. (Reissue edition.)
Fitzgerald, F. Scott. *The Great Gatsby.* Scribner, 1995. (Reprint edition.)
Frederic, Harold. *The Damnation of Theron Ware.* Prometheus Books, 1997.
Gibson, William. *The Miracle Worker.* Bantam Books, 1984.
Gilman, Charlotte Perkins. *Herland.* Pantheon Books, 1979. (Reissue edition.)
Howells, William Deans. *The Rise of Silas Lapham.* New American Library, 1987. (Reissue edition.)
Keyes, Daniel. *Flowers for Algernon.* Skylark, 1984. (Reissue edition.)
Riis, Jacob. *How the Other Half Lives.* Dover, 1971.
Twain, Mark. *The Tragedy of Pudd'nhead Wilson and the Comedy, Those Extraordinary Twins.* Oxford University Press, 1996.
Wells, Ida B. *Crusade for Justice: Autobiography,* ed. by A. Duster. University of Chicago Press, 1970.
________. *Memphis Diary of Ida B. Wells*, ed. by Miriam Decosta-Willis. Beacon Press, 1995.

Reference

Breitbart, Eric. *A World on Display 1904: Photographs from the St. Louis World's Fair.* University of New Mexico Press, 1997.
Daniels, Roger. *Not Like Us: Immigrants And Minorities In America, 1890-1924.* Ivan R. Dee, 1998.
Diner, Steven J. *A Very Different Age: Americans of the Progressive Era.* Hill and Wang, 1998.
Kraut, Alan M. *Silent Travelers: Germs, Genes, and the "Immigrant Menace."* Johns Hopkins University Press, 1999. (Reprint edition.)
Rydell, Robert. *All the World's a Fair.* University of Chicago Press, 1984.
Wertheimer, Barbara Mayer. *We Were There: The Story of Working Women in America.* Pantheon Books, 1977.

Chapter 5: Eugenics and the Power of Testing

Facing History Resources

Facing History and Ourselves: Holocaust and Human Behavior—Chapters 3 and 4

Videos

"Racial Tracking." 60 Minutes CBS, (15 min.)
The Road to Brown California. Newsreel (47 min.)
School: The Story of American Public Education, Episode 2. "As American as Public School, 1900-1950." Films for the Humanities and Sciences (55 min.)
Secrets of the SAT. PBS (60 min.)

Websites

History of Influences in the Development of Intelligence Theory and Testing
http://www.indiana.edu/%7Eintell/index.html
History of Schooling and Social Control
http://www.bc.edu/bc_org/avp/soe/te/pages/docstudwork/ed711/pages/regulator2.html
School: The Story of American Public Education
http://www.pbs.org/kcet/publicschool/about_the_series/index.html

Novels/Memoirs & Autobiographies

Cooper, Michael L. *Indian School: Teaching the White Man's Way.* Houghton Mifflin, 1999.
Lomawaima, K. Tsianina. *They Called it Prairie Light: The Story of Chilocco Indian School.* University of Nebraska Press, 1995. (Reprint edition.)
Shreve, Susan Richards. *Tales Out of School: Contemporary Writers on Their Student Years.* Beacon Press, 2000.
Suskind, Ron. *A Hope in the Unseen: An American Odyssey from the Inner City to the Ivy League.* Broadway Books, 1999. (Reprint edition.)

References

Chapman, Paul Davis. *Schools as Sorters: Lewis M. Terman, Applied Psychology and the American Testing Movement, 1890-1930.* New York University Press, 1990. (Reprint edition.)
Cravens, Hamilton. *The Triumph of Evolution: American Scientists and the Heredity-Environment Controversy.* University of Pennsylvania Press, 1978.
Devlin, Bernie, et al, eds. *Intelligence, Genes, and Success: Scientists Respond to The Bell Curve.* Copernicus Books, 1997.
Herrnstein, Richard J. and Charles Murray. *The Bell Curve: Intelligence and Class Structure in American Life.* Free Press, 1996.
Jacoby, Russell and Naomi Glauberman, eds. *The Bell Curve Debate: History, Documents, Opinion.* Times Books, 1995.
Kamin, Leon. *The Science and Politics of IQ.* Lawrence Erlbaum Associates, 1974.
Trent, James. *Inventing the Feeble Mind: A History of Mental Retardation in the United States.* University of California Press, 1995.
Zenderland, Leila. *Measuring Minds: Henry Herbert Goddard and the Origins of American Intelligence Testing.* Cambridge University Press, 1998.

Chapter 6: Toward Civic Biology

Facing History Resources

Facing History and Ourselves: Holocaust and Human Behavior—Chapters 3 and 4

Videos

The Lynchburg Story: Eugenic Sterilization in America. Filmmakers Library (55 min.)
The Sterilization of Leilani Muir. National Film Board of Canada (47 min.)
Tomorrow's Children. 1934. Sinister Cinema (52 min.)

Websites

Smithsonian National Museum of American History, "The Disability Rights Movement"
http://americanhistory.si.edu/disabilityrights/index.html

Novels/Memoirs & Autobiographies

Larsen, Nella. *Quicksand and Passing.* Rutgers University Press, 1986.

References

Condit, Celeste Michelle. *The Meanings of the Gene: Public Debates about Human Heredity.* University of Wisconsin Press, 1999.
Gallagher, Nancy L. *Breeding Better Vermonters: The Eugenics Project in the Green Mountain State.* University Press of New England, 1999.
McLaren, Angus. *Our Own Master Race: Eugenics in Canada, 1885-1945.* McClelland & Steward Ltd., 1990.
Paul, Diane B. *Controlling Human Heredity: 1865 to the Present.* Humanity Books, 1995.
Pernick, Martin S. *The Black Stork: Eugenics and the Death of "Defective" Babies in American Medicine and Motion Pictures Since 1915.* Oxford University Press, 1996.
Rafter, Nicole Hahn. *Creating Born Criminals.* University of Illinois Press, 1998.
Reilly, Philip R. *The Surgical Solution: A History Of Involuntary Sterilization In The United States.* Johns Hopkins University Press, 1991.
Reverby, Susan M., ed. *Tuskegee's Truths: Rethinking the Tuskegee Syphilis Study.* University of North Carolina Press, 2000.
Trent, James. *Inventing the Feeble Mind: A History of Mental Retardation in the United States.* University of California Press, 1995.

Chapter 7: Eugenics, Citizenship, and Immigration

Facing History Resources

Facing History and Ourselves: Holocaust and Human Behavior—Chapter 6
Guide to America and the Holocaust: Deceit and Indifference

Videos

America and the Holocaust: Deceit and Indifference. PBS (81 min.)

In Search Of Ourselves. PBS (120 min.)
In the Shadow of the Reich: Nazi Medicine. Movies Unlimited (54 min.)
In the White Man's Image. PBS (58 min.)
The Irish in America: Long Journey Home. PBS (4 videos, 6 hrs.)

Websites

Ellis Island
http://www.ellisisland.org *http://www.historychannel.com/ellisisland/index2.html*
Immigrant and Passenger Arrivals: Select Catalog of National Archives Microfilm Publications
http://www.nara.gov/publications/microfilm/immigrant/ipcat.html
United States Immigration and Naturalization Service
http://www.ins.gov/graphics/aboutins/history/teacher/index.htm
University of Minnesota's Immigration History Resource Center
http://www1.umn.edu/ihrc

Novels/Memoirs & Autobiographies

Antin, Mary. *The Promised Land.* Penguin, 1997. (Reprint edition.)
Cahan, Abraham. *The Rise of David Levinsky.* Penguin, 1993. (Reprint edition.)
Danquah, Meri Nana-Ama, ed. *Becoming American: Personal Essays by First Generation Immigrant Women.* Hyperion, 2000.
Galarza, Ernesto. *Barrio Boy.* University of Notre Dame Press, 1971.
Hamill, Pete. *Snow in August.* Little, Brown, 1997.
Hutner, Gordon, ed. *Immigrant Voices: Twenty-four Narratives on Becoming an American.* Signet Classic, 1999.
Inada, Lawson Fusao. *Only What We Could Carry.* Heyday Books, 2000.
Kingston, Maxine Hong. *The Woman Warrior: Memoirs of a Girlhood among Ghosts.* Vintage Books, 1989.
Marshall, Paule. *Brown Girl, Brownstones.* Feminist Press, 1996.
Mazer, Anne. *America Street: a Multicultural Anthology of Stories.* Persea Books, 1993.
Morrison, Joan. *American Mosaic: The Immigrant Experience in the Words of Those Who Lived It.* University of Pittsburgh Press, 1993.
Morrison, Toni. *The Bluest Eye.* Knopf, 2000.
Thomas, Piri. *Down These Mean Streets.* Vintage Books, 1997. (Reprint edition.)

References

Daniels, Roger. *Not Like Us: Immigrants And Minorities In America, 1890-1924.* Ivan R. Dee, 1998.
Diner, Steven J. *A Very Different Age: Americans of the Progressive Era.* Hill and Wang, 1998.
Jacobson, Matthew Frye. *Barbarian Virtues: The United States Encounters Foreign Peoples at Home and Abroad.* Hill and Wang, 2000.

Chapter 8: The Nazi Connection

Facing History Resources

Facing History and Ourselves: Holocaust and Human Behavior—Chapters 6, 7, and 8
Guide to America and the Holocaust: Deceit and Indifference

Videos

America and the Holocaust: Deceit and Indifference. PBS (81 min.)
Black Survivors of the Holocaust. Afro Wisdom Films (62 min.)
Childhood Memories. Facing History and Ourselves (57 min.)
The Democrat and the Dictator. PBS (58 min.)
In the Shadow of the Reich: Nazi Medicine. Movies Unlimited (54 min.)
The People's Century, tapes 4, "Lost Peace, 1919" and 9, "Master Race." PBS (60 min. each)
Rise of the Nazis. Film Library (20 min.)
World War II: The Propaganda Battle. PBS (58 min.)

Websites

Facing History and Ourselves
http://www.facinghistory.org
United States Holocaust Memorial Museum
http://www.ushmm.org

Novels/Memoirs & Autobiographies

Bradbury, Ray. *Fahrenheit 451*. Ballantine Books, 1995. (Reissue edition.)
Claeys, Gregory and Lyman Tower Sarget, eds. *The Utopia Reader.* New York University Press, 1999.
Huxley, Aldous. *Brave New World.* Harper Perennial, 1998 (reprint edition).
Jackson, Shirley. *The Lottery and Other Stories*. Modern Library, 2000.
LeGuin, Ursula K. "The Ones Who Walk Away from Omelas," in *The Wind's Twelve Quarters.* Orion Publishing Group, 2000.
Orwell, George. *1984*. Knopf, 1992. (Everyman Library Edition.)

References

Aly, Götz. *Cleansing The Fatherland: Nazi Medicine And Racial Hygiene*. Johns Hopkins University Press, 1994.
Burleigh, Michael, et al. *The Racial State: Germany, 1933-1945.* Cambridge University Press, 1991.
Burleigh, Michael. *The Third Reich: A New History.* Hill & Wang, 2000.
Kühl, Stefan. *The Nazi Connection: Eugenics, American Racism, and German National Socialism*. Oxford University Press, 1994.
Michalczyk, J. J. *Medicine, Ethics, and the Third Reich: Historical and Contemporary Issues*. Sheed & Ward, 1994.

Proctor, Robert N. *Racial Hygiene: Medicine Under the Nazis*. Harvard University Press, 1989.

Chapter 9: Legacies and Possibilities

Facing History Resources

Facing History and Ourselves: Holocaust and Human Behavior—Chapters 9, 10 and 11
Guide to Choosing to Participate

Videos

Ascent of Man: Knowledge or Certainty. Ambrose Video Publishing (52 min.)
All about Us. ABC News (30 min.)
Eugenics in Sweden: The Secret Shame, The Dark Side of the Social Welfare State. ABC News (22 min.)
Gattaca. TriStar Home Video (107 min.)
Up the Long Ladder. Paramount (46 min.)

Websites

Society for Utopian Studies
http://www.utoronto.ca/utopia/index.html
Utopia: The Search for the Ideal Society in the Western World
http://www.nypl.org/utopia/0_meta.html

Novels/Memoirs & Autobiographies

Lowry, Lois. *The Giver*. Houghton Mifflin, 1993.
______. *Gathering Blue*. Houghton Mifflin, 2001.
Piercy, Marge. *Woman on the Edge of Time*. Crest, 1990. (Reissue edition.)

References

Hubbard, Ruth and Elijah Wald. *Exploding the Gene Myth: How Genetic Information Is Produced and Manipulated by Scientists, Physicians, Employers, Insurance Companies, Educators, and Law Enforcers*. Beacon Press, 1999.
Kaplan, Jonathan Michael. *The Limits and Lies of Human Genetic Research: Dangers for Social Policy*. Routledge, 2000.
Lewontin, R.C., Steven Rose, and Leon J. Kamin. *Not in Our Genes: Biology, Ideology, and Human Nature*. Pantheon Books, 1984. (Out of print, but available in libraries.)
Sacks, Peter. *Standardized Minds: The High Price of America's Testing Culture and What We Can Do to Change It*. Perseus Books, 2001.
Steen, R. Grant. *DNA and Destiny: Nature and Nurture in Human Behavior*. Perseus, 1996.
Suzuki, David and Peter Knudtson. *Genethics: The Clash between the New Genetics and Human Values*. Harvard University Press, 1990.

INDEX